Only One Earth

Forty years after the United Nations Conference on the Human Environment in Stockholm, the goal of sustainable development continues via the Rio+20 conference in 2012. This book will enable a broad readership to understand what has been achieved in the past forty years and what hasn't. It shows the continuing threat of our present way of living to the planet. It looks to the challenges that we face twenty years from the United Nations Conference on Environment and Development, 'The Earth Summit', in Rio, in particular in the areas of economics and governance and the role of stakeholders. It puts forward a set of recommendations that the international community must address now and in the the future. It reminds us of the planetary boundaries we must all live within and and what needs to be addressed in the next twenty years for democracy, equity and fairness to survive. Finally it proposes through the survival agenda a bare minimum of what needs to be done, arguing for a series of absolute minimum policy changes we need to move forward.

Felix Dodds is the Executive Director of Stakeholder Forum for a Sustainable Future. He has been active at the UN since 1990 and set up three global NGO coalitions for UN Conferences, Summits and Commissions, including the UN Commission on Sustainable Development. He is the author or editor of several books, including *How to Lobby at Intergovernmental Meetings* and *Climate Change and Energy Insecurity*.

Michael Strauss is Executive Director of Earth Media, an independent political and communications consultancy based in New York. Michael serves as media consultant and advisor to governments, UN agencies and civil society coalitions. For the Rio+20 Summit in 2012, he was designated Civil Society/Media Liaison by the UN Department of Economic and Social Affairs (DESA). He has served as media coordinator for NGO coalitions since the lead-up to the 1992 'Earth Summit', and has worked extensively with NGOs, Trade Unions, Youth, Women, Indigenous Peoples, and Local Authorities at the UN Commission on Sustainable Development in New York.

Maurice Strong served as Secretary-General of both the United Nations Conference on the Human Environment (Stockholm, 1972), which launched international negotiations on environmental issues and led to the establishment of UNEP, and the United Nations Conference on Environment and Development ('The Earth Summit', 1992, Rio de Janeiro). He was the first Executive Director of the United Nations Environment Programme (UNEP). He has played a unique and critical role in globalizing the environmental movement, as part of an extraordinary career that spanned both business and public service in the fields of international development, the environment, energy and finance. The trail that he pioneered has continued to the Rio+20 conference of 2012.

'For those who follow the story around the international sustainable development agenda, I don't know where we'd be without the publications of Stakeholder Forum. *Only One Earth* is the latest in that series, combining all the front-line insights of Felix Dodds and Michael Strauss with the experience and wisdom of Maurice Strong – a winning combination!'

Jonathon Porritt, Forum for the Future

'Forty years ago Oluf Palme reminded us that we must share and shape our future together – it is a shared responsibility containing difficult choices. A transition towards a green economy is one of those difficult choices. One that requires political leadership. It will not happen unless we make it happen. Let's put the world economies to work for a common, sustainable future – we can't afford otherwise. Dodds, Strauss and Strong provide some suggestions on how we might address these future challenges.'

Ida Auken, Danish Minister for the Environment

'Felix Dodds, Michael Strauss and Maurice Strong use their intimate experience of UN processes to detail the long and sometimes painful journey from the Stockholm summit of 1972 towards Rio+20 in 2012. The distillation of history would be useful to anyone new to the issues. But more important is the dissection of the various forces at play, including trade, competitive development, aid and environmental awareness. Those forces are still here, and will play a major role in shaping the path towards global sustainability – or not – well beyond Rio.'

Richard Black, BBC Environment Correspondent

'It is a privilege to review the recent history of a remarkable initiative that changed attitudes and perceptions, and introduced a new approach for determining the future of Planet Earth. Written by outstanding players that contributed effectively to the success of this major effort, it covers in detail scientific, diplomatic and strategic aspects of a process that peacefully brought together all nations.'

Henrique B. Cavalcanti, Federal Minister of Environment in Brazil. Former Chairman of the UN Commission on Sustainable Development (and Delegate to Stockholm-1972, to Rio de Janeiro-1992, and Johannesburg-2002)

'The first Rio Earth Conference set us on three tracks to sustainable development – social, environmental, economic. In this definitive book, Maurice Strong, one of the architects of that iconic conference, joins with Felix Dodds and Michael Strauss to review the convergence and collisions on the development journey since 1992 and to consider how Rio+20 can truly become a platform for achieving 'the future we want'. Their relevant and enduring message – we are all shareholders in Earth Incorporated and have a responsibility to take the most sustainable paths to prosperity, for people and planet.'

H. Elizabeth Thompson, Assistant Secretary-General, Executive Coordinator UN Conference on Sustainable Development Rio+20 and former Minister for Energy and Environment of Barbados

'At a time of mounting threats to the sustainability of the planet, and as we prepare for the 20-year review of the original Rio Conference, it is essential to understand the road that has brought us to this juncture, so that we can understand the momentous opportunities and tremendous challenges that we face in reaching for a sustainable future. Dodds, Strauss and Strong provide a compelling narrative of the road that brought us to this point and outline a potential path for moving forward.'

Adnan Z. Amin, Director-General, International Renewable Energy Agency

Only One Earth

The long road via Rio to sustainable development

Felix Dodds and
Michael Strauss
with Maurice Strong

Routledge
Taylor & Francis Group

LONDON AND NEW YORK

First published 2012
by Routledge
2 Park Square, Milton Park, Abingdon, Oxon OX14 4RN

Simultaneously published in the USA and Canada
by Routledge
711 Third Avenue, New York, NY 10017

Routledge is an imprint of the Taylor & Francis Group, an informa business

British Library Cataloguing in Publication Data
A catalogue record for this book is available from the British Library

Library of Congress Cataloging in Publication Data
Dodds, Felix.
Only one Earth : the long road via Rio to sustainable development /
Felix Dodds and Michael Strauss with Maurice Strong.
    p. cm.
  "Simultaneously published in the USA and Canada"–T.p. verso.
  Includes bibliographical references and index.
  1. Sustainable development–History–20th century. 2. Sustainable
development–History–21st century. 3. United Nations Conference
on Environment and Development (1992 : Rio de Janeiro, Brazil)–
Influence. 4. Nature–Effect of human beings on. 5. Environmental
protection. 6. Sustainable development–Citizen participation.
7. Environmental policy. 8. Economic policy. I. Strauss, Michael,
1952– II. Strong, Maurice F. III. Title.
HC79.E5D623 2012
338.9′27–dc23   2012002358

ISBN: 978-0-415-54025-4 hbk
ISBN: 978-0-203-10743-0 ebk

Typeset in Bembo
by HWA Text and Data Management, London

Contents

Foreword

Maurice Strong

This impressive and timely book illuminates thoughtfully and authoritatively the long road to Rio+20. In establishing the broader context of this landmark event it makes a uniquely important contribution to it. I am pleased at the opportunity I have had to participate in its preparation.

There have been immense changes in the world since the first UN Conference on the Human Environment in Stockholm in 1972 put the environment on the world's agenda. Some notable progress has since been made in awareness and understanding of the issues we must address, in our capacity to do so, in the urgency of the need for decisive actions and the dire consequences of our failure to act.

The ominous paradox is that the will to act suffers today from a decline in public attention and the preoccupation of governments and people with more immediate financial and economic concerns. This is reflected in the continued lack of progress in implementing past commitments as well as the prospect of undertaking new ones at Rio+20. This recession in political will will have far more damaging consequences for the human future than the more immediate issues that give rise to it.

Indeed, it has never been more important to heed the evidence of science that time is running out in our ability to manage successfully our impacts on the Earth's environment, biodiversity, resource and life-support systems on which human life as we know it depends. We must rise above the lesser concerns that pre-empt our attention and respond to the reality that the future of human life on Earth depends on what we do, or fail to do in this generation. What we have come to accept as normal is not normal as increased human numbers, the growing intensity of human impacts and the demographic dilemma faced by so many nations are returning the Earth to the conditions that have existed for most of its existence, that do not support human life as we know it.

We must deal with this as the most dangerous security issue humanity has ever faced, threatening the very conditions necessary for the support and sustainability of life on Earth.

It is in this larger context that we must view Rio+20 as a unique opportunity to make the 'change of course' called for by business leaders at the Earth Summit in 1992. It requires fundamental changes in the way in which we manage the activities through which we impact on the Earth's sustainability. This will require a degree of cooperation beyond anything we have yet experienced at a time when competition and conflict over scarce resources is escalating.

The decisions and policies which determine our impacts on sustainability are primarily motivated by economic and financial considerations. The transcendent importance of the actions to be taken at Rio+20 requires that they be firmly rooted in our deepest moral and ethical principles. This is why I feel so strongly that Rio+20 must endorse and be grounded in the Earth Charter. The change of course called for at Rio in 1992 requires radical changes in our current economic system, particularly by those countries, mostly Western, which have dominated the world economy during the past century and will be most resistant to change. Yet they have monopolized the economic benefits that have accompanied our cumulative damage to the Earth's life-support systems, its precious biological resources and its climate.

Rio+20 must reinforce the focus on biodiversity to which this Decade on Biodiversity is devoted so that it will lead to specific actions on implementation of the measures required to protect these resources so essential to global sustainability.

Experience has demonstrated that those countries that have been most successful in improving their environment are those, like Japan, which have been most efficient in managing their economies and reducing the energy, resources and materials used to produce their GDP. Rio+20 must provide for special measures to assist developing countries in the efficiency of their economies.

No issue is more important to the human future than that of climate change, in which the political will to act cooperatively and decisively has dangerously diminished. Rio+20 must reinforce international efforts taken at Durban and beyond to reach agreement and renewal of the Climate Change Convention and its implementation. Paradoxically, if we fail to act, the reduction in global greenhouse gas emissions could occur through the collapse of the world economy, to which none of us would aspire. After all, the roots of the environmental and climate change crises are the same as those of the economic and financial crises – the inadequacies of our current system.

Only an enlightened view of their self-interest in the security and sustainability of life is likely to induce the more developed countries to accept the principal responsibility they bear for the fundamental change of course that we must make. Developing countries must play their part but their responsibilities are of a different order of magnitude. The concept of shared but differentiated responsibilities must be strongly reinforced at Rio+20.

The growing inequities in sharing the benefits of economic growth continue to provide a widening rich–poor divide in both developing and more developed countries, even in China, which has lifted more people out of poverty than any nation has ever done. This undermines the prospects of enabling the poor and disadvantaged to share fully and equitably in the benefits of sustainable development and will lead to social unrest, evidence of which is already emerging.

This book will help to create much greater public interest and awareness in Rio+20 as an event in itself. This is what helped to attract unprecedented numbers of world leaders, media and non-governmental organizations to the Earth Summit in 1992 and move governments to take decisions beyond original expectations. But time is short, and the resources available to the secretariat and others preparing for Rio+20 are meagre.

However much can still be done to give greater visibility, public awareness and political priority to Rio+20 by making the event itself more attractive. This would enhance the awareness of and public attention to Rio+20 and provide further incentives for world leaders to participate.

The United Nations alone cannot be expected to take the main role in this but it can reach out to the many others around the world that are willing and able to engage. The fact that UN Secretary-General Ban Ki-moon has accorded Rio+20 and its follow-up highest priority will help create a positive response by others. The work of the preparatory committee and those who are already engaged can help to mobilize such wide spread support and ensure that the conference produces the results on which the security and sustainability of life on Earth depends.

A particularly distinctive feature of this book is its focus on the important role of 'stakeholders': those important actors which are not governments. Never has their role been more important than it is now as they take the lead in mobilizing the support and participation of civil society in the conference. The fact that the authors are themselves leading stakeholders deeply engaged in supporting preparations for Rio+20 makes their views as expressed in this book particularly authoritative.

This also gives greater credence to the many practical recommendations they make in the book. It underscores the key role stakeholders will have in implementation by governments.

In identifying the origins, the progress and the pitfalls of the long road to Rio+20 the authors of this book have made a unique and indispensable contribution to its prospects of success.

Introduction

Felix Dodds and Michael Strauss

Twelve years into a new millennium, it doesn't look as if we or our planet are doing very well. Extreme and erratic weather events (and can anyone *really* claim with a straight face that they're not exacerbated by global warming?) have progressed from the merely atypical to the virtually incredible. Indices of the environmental quality of other critical areas – oceans, biodiversity, air quality, freshwater, forests – continue to race to the bottom of their ranges.

Most developed countries' economies are mired deep in near-catastrophic recession – although their governments refuse to reject the wholesale de-regulation of the finance industry that brought them to the edge. Efforts to fulfil the Millennium Development Goals (MDGs) – which, after all, only attempt to reach half of the billion people living below survival level – are at best only partly there.

And the level of willing cooperation among peoples and among nations – the cornerstone of progress toward resolving complex, multi-sectoral problems – sometimes feels as though it's run into a bottomless hole in the ground.

In considering the trajectory of the last 40 years of environmental, social and economic actions – and the challenges that still loom ahead – it would be easy to veer towards pessimism.

And yet … there are repeated, refreshing reminders that there *is* cause for hope. There are abundant models of liveable communities, of integrated renewable energy systems, of non-polluting transportation and manufacturing technologies, and of organic agriculture that evoke shimmering glimpses of an exciting, attractive *and achievable* world that would be a glorious place to live. Most people in all countries want to live in such a world. Most parents want such a future for their families. Most youth and children know instinctively that there is a better way possible – and are ready to work for it.

What they, and we, often lack is a vision – a positive vision – of how such a world would look and how it could be achieved. It is up to us to provide that vision and to show in a realistic way the path or paths for getting there.

Because these past 20 years have seen considerable successes. The negotiation of intergovernmental treaties on biodiversity, desertification and climate – no matter how obstructed. The increasingly system-wide opening of UN processes to NGOs, to workers' organizations, to indigenous peoples, to local governments and, yes, to business. And there can be many more. What has been blocking them is a distinct lack of political will at the highest level to implement the major agreements on environment, on the MDGs, and on sustainable development – agreements that governments have already signed.

We are at a critical time, one that needs enlightened leadership that builds on our better selves. We are in a period with only one major super power, the United States. Yet the US seems unable to show leadership on climate change, on biodiversity, on energy and sustainable economic transformation, or even on its own domestic environmental resources which are under threat, as is even the existence of its Environmental Protection Agency. William Reilly, the former EPA Director under President George H.W. Bush, expressed his dismay recently when he said that for "some of the most prominent leaders of the Republican Party, science has left the building" (Reilly, 2011).

The moral and intellectual energy and political will-power to redress these failures has been catalysed before, and it can be mobilized again.

In 1972, under commission by the Secretary-General of the approaching World Conference on the Human Environment in Stockholm, Barbara Ward and René Dubos authored a report called *Only One Earth*. It sounded an impassioned call to arms, an urgent alarm about the impact of human activity on the planet. The authors maintained a tone of optimism that humans could together address the challenges we all faced. The slogan, 'Only One Earth', became the rallying cry of Stockholm. And – as relevant today as it was then – is the title we take for this book.

Today, 'Only One Earth' attempts to help re-mobilize that energy among people – and re-build that political will – by reminding us of how we got here in the recent past, and suggesting some of the directions in which we need to go.

Part I of this book takes the reader from the preparations for the Stockholm Conference in 1972 to preparations for the World Conference on Sustainable Development in 2012. As we write this, we know that 2012 will not be like Stockholm nor like the 1992 Earth Summit in Rio. It will not be a place of agreement on major conventions and action plans, but more a launching pad for initiatives, for sustainable development goals and perhaps for future conventions. We hope it will mark the re-emergence of sustainable development as the key organizing framework for how people live on this planet.

It's clear that the past provides us with lessons for the future. As always, though, the question is, will we learn them?

Part II tries to describe some of the frustrating roadblocks to implementing what was agreed in Rio and Johannesburg, and to identify achievable strategies and pathways for moving around them. Its chapters look at the implementation gap, the democracy gap, and the governance gap between promises and actions, and the multiple challenges of transforming the economy in a sustainable direction. It then presents a 'survival agenda', with specific, realizable actions that can help nations, economies and people to move onto a sustainable path.

Nations face numerous challenges as they start to define the second decade of the third millennium. The multilateral system that governments have created for environment and sustainable development is weak and fragmented compared to those established for trade and finance. If we are to secure a fair and equitable world for all people, this must change. There is no reason that the same sense of creative compromise and necessary cooperation cannot pertain.

Our democracies are fragile, as we have recently realized, as politicians fail to live up to even their own professed values, and public support for them drops to the lowest levels on record. At the same time, seemingly out of nowhere, a courageous and vibrant people's movement has emerged in the Middle East, in Russia and even in the US and elsewhere that transcends culture and governments. It offers a hope for a re-invigoration of democracy and a renewal of idealism. The recognition that governments need to govern with the support of key stakeholders – and all the people – has become clearer since Rio. We are starting on a path to towards participatory democracy at the same time as the threats that surround us are increasing.

Our economies have been built on unsustainable consumption patterns. We challenge completely the soundbite philosophy of US President George H.W. Bush in 1992 who said that 'the American Way of Life is non-negotiable'. One may choose to negotiate or not to negotiate. But the sad irony here is that by not doing so, that same country has brought the risk and reality of degradation to its 'way of life' far closer.

Instead of embracing and defining a new direction, and providing the resources for developing countries to choose a more sustainable way of developing, much of the last 20 years has been wasted. China and India have eagerly adopted the consumption patterns of the West and now China is the biggest absolute carbon dioxide producer. So the challenges are now much greater than they might have been.

The question then is do we 'together' address the challenges of the emerging agenda of environmental security, or do we fall back on selfishness and fear? The kind of world that could emerge in the second scenario will be far less democratic, less fair, less hopeful and less sustainable. Perhaps US Senator Robert F. Kennedy summed up the stakes of our future challenges best:

> For when you teach a man to hate and to fear his brother, when you teach that he is a lesser man because of his color, or his beliefs or the policies that he pursues, when you teach that those who differ from you, threaten your freedom or your job or your home or your family, then you also learn to confront others not as fellow citizens, but as enemies. To be met not with cooperation but with conquest, to be subjugated, and to be mastered. We learn at the last to look at our brothers as aliens. Alien men with whom we share a city, but not a community. We learn to share only a common fear, only a common desire to retreat from each other, only a common impulse to meet disagreement with force. Our lives on this planet are too short, the work to be done is too great. But [if] we can perhaps remember, that those who live with us are our brothers, that they share with us the same short moment of life that they seek as do we, nothing but the chance to live out their lives in purpose and in happiness, surely this bond

of common fate, this bond of common roles can begin to teach us something, that we can begin to work a little harder, to become in our hearts brothers and countrymen once again.

(Kennedy, 1968a)

As for ourselves, we are unabashed multi-lateralists and supporters of the United Nations. We believe governments need to adequately fund the UN – a body that, however imperfectly, has been committed to saving succeeding generations from the scourge of war for over 65 years, that promotes the equal rights of men and women and of nations large and small, and that has empowered social progress, freedom and better standards of life for hundreds of millions of people on every continent.

The defining agenda of the last few years has been that of the financial crisis – a crisis caused by excessive short term thinking and by outright greed. The need to reform the global economic system to address both its own stability and the sustainable development of all countries could not be more apparent. Faced with a need for financial stimulus to keep national economies from collapsing, some countries have focused their recovery packages on green jobs and investment (South Korea and China are among the few examples). Others have focused exclusively on regulation without investment (as in the EU), or investment without regulation (the effective reality in the US) – propping up the very individuals and institutions who had caused the problem.

But few countries tried to integrate the fundamental principles of ecosystem and resource limits into their existing economic plans. They should have – the crisis presented an unusual opportunity to. As Anil Agarwal (1992) put it: 'Do not worry about the Gross National Product; preserve and promote the Gross Natural Product'.

As governments move towards the UN's Millennium Development Summit in 2015 we will see how far what was started in Rio in 2012 has gone to reshape the economic agenda. We will see if we have provided the intellectual and political stimulus to reinvigorate multilateral cooperation, which remains the only answer to many of the most intractable problems facing the world today. And we will see if we have captivated the energies and enthusiasms of the next generation to become the agents for change that their world so desperately requires.

We do believe this is a turning point, a time of hope and more importantly a time of sharing this one and only earth among all of its varied passengers. As Adlai Stevenson, a candidate who was criticized as being 'too intelligent' to be elected to the presidency of a major nation, once said in his role, instead, as Ambassador to the UN:

We travel together, passengers on a little spaceship, dependent on its vulnerable reserves of air and soil, all committed, for our safety, to its security and peace. Preserved from annihilation only by the care, the work and the love we give our fragile craft.

(Stevenson, 1965)

We would like to thank some of the people who have made this book possible with their advice, their support and their significant input – Tim Hardwick and our friends at Earthscan. Derek Osborn, Eela Dubey, Farooq Ullah, Georgie Macdonald, Jack Cornforth, Jeannet Lingan, Richard Sherman and Kirsty Schneeberger at Stakeholder Forum. Megan Howell and Sheila Shettle, who, in an earlier age, helped build the foundations of their respective organisations to the levels they are today. And Jan-Gustav Strandenaes, who epitomizes the best of being an NGO and has always been there as both a colleague and a friend.

We want to cite some of the individuals along the 20-year road from the original Earth Summit at Rio who at critical points have provided invaluable dedication and wisdom to making multi-stakeholder cooperation work on the international stage – Barbara Bramble, Clif Curtis, Peter Padbury, Martin Khor, Chee Yoke Ling, Vicky Corpuz, Tom Goldtooth, Saradha Ayer, Neth Dano, Pauulu Kamarakafego, Federica Pietracci and Chantal Line Carpentier, to name only a few.

We also would like to thank for their specific help with this book – Paul Clements Hunt, for his substantive input on the finance, capital markets and the green economy sections, David Taylor, Rachel Kyte, Maria Figueroa Kupcu, Bill Mankin, Mohammed El-Ashry, Hannah Stoddart, Bedrich Moldan, Margaret Brusasco-Mackenzie, Sondra Sullivan and David Le Blanc.

In particular, we'd like to acknowledge Maurice Strong – former entrepreneur, energy company CEO, NGO, Executive Director of UNEP, Secretary-General of the UN conferences at Stockholm *and* at Rio, author, president of the council of the UN University for Peace, and professor at Peking University – who in some ways can be said to have single-handedly invented the intergovernmental environmental and sustainable development governance process. Besides the obvious professional debt, our personal thanks for the lessons learned from years of observing – up close or from a distance, sometimes as rivals but usually as friends, always with awe and admiration.

Since the original Earth Summit at Rio, we have lost some of the leading champions who helped create the sustainability movement from both in front of and behind the scenes. This book is dedicated to them: Joke Waller-Hunter, Michael McCoy, Bella Absuz, Chip Lindner, Wangari Maathai, Svend Auken, Peter Thacher, Richard Sandbrook, Ken Saro Wiwa, Anil Agarwal, Maximo Kalaw and Chico Mendes. Their contributions have helped us, inspired us and given us hope that we can move towards a more sustainable future.

As Vaclav Havel, another peaceful warrior who left us in 2011, put it in two comments that remind us what it takes to lead a benevolent revolution:

> None of us – as an individual – can save the world as a whole, but ... each of us must behave as though it were in his (or her) power to do so.

> May truth and love triumph over lies and hatred.
>
> (Havel, 1994)

Abbreviations

10YFP	10-Year Framework of Programmes on Sustainable Consumption and Production
ACC	Administrative Committee on Coordination (United Nations)
ANPED	Northern Alliance for Sustainability
AOSIS	Alliance of Small Island States
BASD	Business Action for Sustainable Development
BASIC	Brazil, South Africa, India and China
BAU	business as usual
BCSD	Business Council for Sustainable Development
BRIC	Brazil, Russia, India and China
CBD	Convention on Biological Diversity
CCAMLR	Convention for the Conservation of Antarctic Marine Living Resources
CCD	Convention to Combat Desertification
CCIC	Canadian Council for International Cooperation
CDM	Clean Development Mechanism
CEB	Chief Executive Board
CER	Certified Emission Reduction
CFC	Common Fund for Commodities
CFCs	chlorofluorocarbons
CGG	Commission on Global Governance
CHS	*see* UNCHS
CIDA	Canadian International Development Agency
CIO-ALU	Congress of Independent Organizations – Associated Labour Unions
CITES	Convention on International Trade in Endangered Species of Wild Flora and Fauna
CMS	Convention on Migratory Species
COCF	Centre for Our Common Future
COP	Conference of Parties (of the UN FCCC)
CSD	*see* UNCSD
CSO	civil society organizations

DDA	Doha Development Agenda
DDT	dichlorodiphenyltrichloroethane
DESA	*see* UNDESA
DPCSD	Department for Policy Coordination and Sustainable Development (United Nations)
DREAMS	development reconciling environment and material success
DSD	Division on Sustainable Development (United Nations)
ECLAC	Economic Council for Latin America and the Caribbean
ECOSOC	Economic and Social Council (United Nations)
EIA	Environmental Investigation Agency
ELCI	Environmental Liaison Centre International
EMG	Environment Management Group
EPA	Environmental Protection Agency
ESG	environmental, social and governance
FAO	Food and Agriculture Organization (United Nations)
FCCC	*see* UNFCCC
FDI	foreign direct investment
FTT	financial transaction tax
G77	group of 77 developing countries
GA	General Assembly (United Nations)
GAPP	Generally Accepted Principles and Practices
GATT	General Agreement on Tariffs and Trade
GDP	gross domestic product
GEC	Green Economy Coalition
GEF	[used in 1997, not defined there]
GEO	Global Environmental Outlook
GEOSS	Global Earth Observation System of Systems
GMEF	Global Ministerial Environmental Forum (UNEP)
GNP	gross national product
GPA	Global Programme of Action
HCFCs	hydrofluorocarbons
IACSD	Inter-Agency Committee on Sustainable Development
ICC	International Chamber of Commerce
ICE	International Court for the Environment
ICFTU	International Confederation of Free Trade Unions
ICJ	International Court of Justice
ICLEI	International Council for Local Environmental Initiatives
ICSU	International Council of Scientific Unions
ICZM	integrated coastal zone management
IEG	international environmental governance
IFC	International Facilitating Committee
IFF	Intergovernmental Forum on Forests (United Nations)
IFI	International Financial Institution
IFSD	institutional framework for sustainable development

IGO	intergovernmental organisation
ILO	International Labour Organization
IMF	International Monetary Fund
IMO	International Maritime Organization
IPBES	Intergovernmental Panel of Biodiversity and Ecosystem Services
IPCC	Intergovernmental Panel on Climate Change
IPF	Intergovernmental Panel on Forests
IPRs	Intellectual Property Rights
IPSD	Intergovernmental Panel on Sustainable Development
ITFF	Interagency Task Force on Forests
ITTA	International Tropical Timber Agreement
ITTO	International Tropical Timber Organization
IUCN	International Union for Conservation of Nature
JCL	Johannesburg Climate Legacy
JPoI	Johannesburg Plan of Implementation
LA21	Local Agenda 21
LDC	least-developed country
MDG	Millennium Development Goal
MEA	multilateral environmental agreement
MOI	means of implementation
MSP	multi-stakeholder processes
NAM	'non-aligned' movement
NCSD	national councils on sustainable development
NGLS	Non-Governmental Liaison Service (United Nations)
NGO	non-governmental organization
NIEO	new international economic order
NRG4SD	Network for Regional Government for Sustainable Development
NSSD	National Strategies for Sustainable Development
NWF	National Wildlife Federation
ODA	Official Development Aid
OECD	Organisation for Economic Co-operation and Development
OPEC	Organization of the Petroleum Exporting Countries
OSPAR	Convention for the Protection of the Marine Environment of the North-east Atlantic
PIC	prior informed consent
POP	persistent organic pollutant
PrepCom	Preparatory Committee (for UN Conferences and Summits)
PRI	Principles for Responsive Investment
PRS	poverty reduction strategy
PRTR	pollutant release and transfer registers
PSI	Principles for Sustainable Insurance
RIM	Regional Implementation Meeting
SAICM	Strategic Approach to International Chemicals Management
SARD	Sustainable Agriculture and Rural Development

SCP	sustainable consumption and production
SDG	sustainable development goals
SEEA	System of Environmental-Economic Accounts
SEED	Supporting Entrepreneurs for Sustainable Development
SIDS	Small Island Developing States
SISD	Summit Institute for Sustainable Development
SIWI	Stockholm International Water Institute
SLUDGE	slightly less unsustainable development genuflecting to the environment
SMEs	small- and medium-sized enterprises
SWF	sovereign wealth fund
TAI	The Access Initiative (World Resources Institute)
TNC	transnational company
UCLG	United Cities and Local Government
UN	United Nations
UNAIDS	Joint United Nations Programme on HIV/Aids
UNCCD	United Nations Convention to Combat Desertification
UNCED	United Nations Conference on Environment and Development
UNCHS	United Nations Centre for Human Settlements
UNCLOS	UN Convention on the Law of the Sea
UN CSD	UN Commission on Sustainable Development
UNCSD	UN Conference on Sustainable Development
UNCTAD	United Nations Conference on Trade and Development
UNCTC	UN Centre for Transnational Corporations
UNDESA	United Nations Department of Economic and Social Affairs
UNDP	United Nations Development Programme
UNECE	United Nations Economic Council for Europe
UNEO	United Nations Environment Organization
UNEP	United Nations Environment Programme
UNEP GC	United Nations Environment Programme Governing Council
UNESCO	United Nations Educational, Scientific and Cultural Organization
UNFCCC	United Nations Framework Convention on Climate Change
UNFI	UNEP Finance Initiative
UNFPA	United Nations Population Fund
UNGASS	United Nations General Assembly Special Session to Review and Appraise the Implementation of Agenda 21
UNIDO	United Nations Industrial Development Organisation
US EPA	US Environmental Protection Agency
WACC	Water and Climate Coalition
WBCSD	World Business Council for Sustainable Development
WCED	World Commission on Environment and Development
WEDO	Women's Environment and Development Organization
WEHAB	Water, Energy, Health, Agriculture and Biodiversity
WEO	World Environmental Organization

WHO	World Health Organization
WIPO	World Intellectual Property Organization
WRI	World Resources Institute
WSSD	World Summit on Sustainable Development
WTO	World Trade Organization
WTTC	World Travel and Tourism Council
WWF	World Wide Fund for Nature
ZPG	zero population growth

About the authors

Felix Dodds is the Executive Director of Stakeholder Forum for a Sustainable Future. He has been active at the UN since 1990, attending the World Summits since the Rio Earth Summit.

He has set up three global NGO coalitions for UN Conferences, Summits and Commissions. These are the UN Commission on Sustainable Development (1993), the UN Habitat II (1995) the WHO Health and Environment Conference (1999). He co-chaired the NGO Coalition at the UN Commission on Sustainable Development from 1997 to 2001. He introduced Stakeholder Dialogues in 1996 through the UN General Assembly in November 1996 for Rio+5 and helped run some of the most successful ones at Bonn Water (2001) and Bonn Energy (2004).

From 1985–1987 he was the chair of the UK Liberal Party's youth wing credited as a major influence on greening the party. From 1997–2001 he co-chaired the UN Commission on Sustainable Development NGO Steering Committee.

He has written or edited the following books: *Biodiversity and Ecosystem Insecurity* (April 2011); *Climate Change and Energy Insecurity* (2009); *Negotiating and Implementing Multilateral Environmental Agreements* (2007); *Human and Environmental Security: An Agenda for Change* (2005); *How to Lobby at Intergovernmental Meetings: Mine is a Café Latte* (2004); *Earth Summit 2002: A New Deal* (2000); *The Way Forward: Beyond Agenda 21* (1997); *Into the Twenty-First Century: An Agenda for Political Realignment* (1988).

He is also a regular contributor to the BBC web site and enjoys blogging from Film Festivals. Most recently he has been/is on the Advisory Boards for: the Collaborative Institute for Climate Oceans and Security, Eye of the Earth Steering Committee (2011), The Bonn 2011 Water Energy and Food Security Nexus Conference, Planet under Pressure: new knowledge towards solutions 2012 global science Conference (2012).

He chaired the 64th United Nations Department of Public Information NGO Conference, titled 'Sustainable Societies – Responsive Citizens' (September 2011).

He was mentioned in an article posted in Associated Degree as one of '25 environmentalists ahead of their time' (2010). He has been advisor to the UK and Danish governments and the European Commission.

Michael Strauss is Executive Director of Earth Media, an independent political and communications consultancy based in New York.

He serves as media consultant and advisor to governments, UN agencies and civil society coalitions. For the Rio+20 Summit in 2012, he was designated Civil Society/Media Liaison by the UN Department of Economic and Social Affairs (DESA). He has served as media coordinator for NGO coalitions since the lead-up to the 1992 Earth Summit, and has worked extensively with NGOs, trade unions, youth, women, indigenous peoples, and local authorities at the UN Commission on Sustainable Development in New York.

International NGOs and coalitions he has advised include Greenpeace International, Friends of the Earth, Sierra Club, WWF, Earth Day Network, IUCN, Consumers International, Oxfam International, WSPA, and the Ocean Noise Coalition. He has coordinated press conferences for the Alliance of Small Island States (AOSIS) – the governmental coalition of 40 island nations at the United Nations – on the issue of climate change. He has worked with both green and responsible business associations, and with international and national labour unions. In Johannesburg, in 2002, at the World Summit on Sustainable Development – at the request of the UN Department of Economic and Social Affairs (DESA) and Department of Public Information (DPI) – he served as organizer and moderator of the daily press conferences by representatives of NGOs, trade unions, youth, women and indigenous peoples. News coverage generated by those press conferences included hundreds of articles and broadcast reports citing the positions of civil society speakers.

He also provided daily commentary on South Africa Broadcasting Company's direct radio coverage. And he conducted daily media workshops for youth NGOs. News coverage he helped generate has appeared in *The New York Times*, BBC, CNN, AP, Reuters, Agence France-Presse, ARD, *O Globo* (Brazil), *Asahi Shimbun* (Japan), SABC (South Africa), TF1 (France), ABC (Australia), CBC (Canada), *The Independent*, *The Hindu* (India), *Politiken* (Denmark), and many others.

He has co-authored two previous books on policy and media advocacy: *Negotiating and Implementing Multilateral Environmental Agreements – A Manual for NGOs* (UNEP Division of Environmental Law and Conventions, 2007) and *How to Lobby at Intergovernmental Meetings* (Earthscan, 2004). He edited *The Dialogue Records* (Northern Clearing House, New York, 1999), the first account of the UN government–civil society dialogue sessions on business and sustainable development. He has guest lectured at graduate seminars at Columbia University's School of International Policy (SIPA) and School of Journalism, and at the New School University Graduate Program in International Affairs. He conducts frequent Media Workshops for NGOs and youth organizations. He is a contributing editor to *Outreach*, the daily newsletter of civil society organizations at UN negotiations, and is a frequent editor of NGOs' written statements and an adviser and political strategist on political issues and negotiations. His typical habitats include the UN's Vienna Café, the offices of UN journalists, and New York's Second Avenue diners.

Maurice Strong has played a unique and critical role in globalizing the environmental movement, as part of an extraordinary career that spanned both business and public service in the fields of international development, the environment, energy and finance.

For the United Nations, he has served as Secretary General of both the United Nations Conference on the Human Environment (Stockholm, 1972), which launched international negotiations on environmental issues and led to the establishment of the United Nations Environment Programme (UNEP), and the United Nations Conference on Environment and Development ('The Earth Summit', Rio de Janeiro, 1992). He was founding Executive Director of UNEP. He was personal envoy of the Secretary-General to lead international support in humanitarian and development needs of the Democratic People's Republic of Korea.

In the world of business, he has been President of Ajax Petroleum; President of PetroCanada, Chairman and CEO of Ontario Hydro (then North America's largest electric power utility), President of PowerCorporation, and Chairman of AZL Resources Incorporated. He has been a Director of the World Economic Forum. He has been a Senior Advisor to the President of the World Bank, and a Member of the International Advisory of Toyota Motor Corporation.

Mr Strong served as Deputy Minister of Foreign Affairs for Canada, head of the Canadian International Development Agency (CIDA) and Chairman of the Canada Development Investment Corporation.

In the NGO world, he has served as Chairman of the World Resources Institute, President of the Earth Council, Canadian President of the World Alliance of YMCAs, and a board member of the World Wildlife Fund, the World Business Council for Sustainable Development, the African-American Institute, the Eisenhower Fellowships, the International Institute for Sustainable Development, the Stockholm Environment Institute and the World Conservation Union.

In the academic world, he has served as President of the University of Peace, Honorary Professor, Institute for Research on Security and Sustainability for Northeast Asia (and Honorary Chairman of its Environmental Foundation, and Chair of its Advisory Board), and sat on the Advisory Council for the Center for International Development at Harvard University.

Former United Nations Secretary-General Kofi Annan paid him the following tribute:

> Looking back on our time together, we have shared many trials and tribulations and I am grateful that I had the benefit of your global vision and wise counsel on many critical issues, not least the delicate question of the Korean Peninsula and China's changing role in the world. Your unwavering commitment to the environment, multilateralism and peaceful resolution of conflicts is especially appreciated.

Part I

The journey from Stockholm

Chapter 1

How we arrived here

The achievements of Stockholm

Our purpose here is to reconcile man's legitimate, immediate ambitions with the rights of others, with respect for all life supporting systems, and with the rights of generations yet unborn. Our purpose is the enrichment of mankind in every sense, of that phrase. We wish to advance – not recklessly, ignorantly, selfishly and perilously, as we have done in the past – but with greater understanding, wisdom and vision. We are anxious and rightly so, to eliminate poverty, hunger, disease, racial prejudice and the glaring economic inequalities between human beings.

Opening Statement, United Nations Conference on the
Human Environment, 4 June 1972 (Strong, 1972)

The 1960s constituted in some ways the most atypical decade in a century filled with unprecedented revolution, war, destruction, economic depression and ultimately affluence. Tens of thousands of people in dozens of countries, mostly young, poured into the streets to demand political and economic freedoms for others – or walked away from the mainstream to look for new cultural freedom for themselves.

It was a start of new movements in favour of equal opportunity, civil rights and feminism – and action against racism and war. At the end of the decade, a new idea began to attract public attention – a movement to protect the environment.

The decade also started with the publication of the path-setting work *Silent Spring* by Rachel Carson. The book highlighted the environmental problems caused by synthetic pesticides. It eventually led to the US ban on dichlorodiphenyltrichloroethane (DDT) and other pesticides. And it initiated the beginning of the environmental movement.

One of the key concerns in the 1960s for the environmental movement and others was population. Between 1900 and 1960 the number of people on the planet had doubled from 1.61 billion to 3.02 billion and by 1968 it had grown to 3.5 billion. Kingsley Davis, an American sociologist and demographer, coined the term 'zero population growth', leading to a ZPG movement that called for population stabilization.

This was an early acknowledgement of the limits to the earth's carrying capacity. Today, with world population at 7 billion, the issue is as relevant as ever. The global fertility rate is still high at 2.5 (2010); this is still above the 'replacement level' of 2.1 children per woman. The present estimate is that the world's population will grow to peak at 11 billion by 2050.

The most turbulent year of the 1960s was 1968. It started with a surge of hope with the election in Czechoslovakia of Alexander Dubček (5 January) as leader of the Communist Party. His reforms would lead to the peaceful, populist uprising dubbed Prague Spring, before being crushed by 200,000 Warsaw Pact troops re-imposing authoritarian Soviet rule (21–22 August). The year would see student and youth demonstrations across the world seeking a more participatory democracy – uprisings that would be echoed, astonishingly, 20 years later with the fall of the Berlin Wall and nearly 40 years later with the Arab Spring.

Paris was immobilized for weeks by thousands of striking students and workers protesting militarism and demanding a more egalitarian economy. Canada elected as prime minister its most charismatic leader, Pierre Trudeau (20 April 1968), who refused to support the US war in Vietnam, acknowledged he'd smoked marijuana, and called for a 'Just Society'.

But 1968 would also see the assassination in the US of two inspiring leaders of that potential new world – Martin Luther King (4 April) and Robert F. Kennedy (5 June). They had offered an agenda of hope not only to a generation of young people in their own country, but to millions across the world who had felt oppressed, deprived or excluded.

In a telling preview of the agenda at the centre of the 2012 Earth Summit, Robert Kennedy 40 years earlier had criticized an over-reliance on the prevailing measure of material production – the gross national product (GNP) – as being an erroneous indicator of a country's growth and wellbeing.

> It [GNP] counts air pollution and cigarette advertising, and ambulances to clear our highways of carnage. It counts special locks for our doors and the jails for the people who break them. It counts the destruction of the redwood and the loss of our natural wonder in chaotic sprawl. It counts napalm and counts nuclear warheads and armored cars for the police to fight the riots in our cities. It counts [the assassin's] rifle and [the murderer's] knife, and the television programs which glorify violence in order to sell toys to our children. Yet the gross national product does not allow for the health of our children, the quality of their education or the joy of their play. It does not include the beauty of our poetry or the strength of our marriages, the intelligence of our public debate or the integrity of our public officials. It measures neither our wit nor our courage, neither our wisdom nor our learning, neither our compassion nor our devotion to our country, it measures everything in short, except that which makes life worthwhile.
>
> (Kennedy, 1968b)

The UK Prime Minister, David Cameron, called for something similar when he declared that there was 'more to life than money' (21 November 2010). He was referring to the use of the current index of economic prosperity measure, the gross domestic product (GDP). Cameron's coalition government of Liberals and Conservatives has argued that prosperity should not be solely measured using economic indicators, and suggested a shift to a 'happiness survey' as promoted by Nobel prize-winning economists Joseph Stiglitz and Amartya Sen.

Little noted in most reviews of 1968, but as important as many of the key moments mentioned above, was the Swedish government's introducing a resolution in the UN General Assembly to convene a world conference on the environment. The motion passed in 1969, when it was agreed to hold the UN Conference on Human Environment in 1972, in Stockholm.

The 1960s closed with the Apollo 11 landing on the moon (20 July 1969) and its iconic photo of the Earth, taken from lunar orbit. For the first time human beings all saw their home planet from outer space.

It is perhaps difficult to remember the powerful – and complex – impact that the Apollo 11 landing had on people all over the world. On one hand, it represented a massive technological achievement – perhaps the culmination of the science-as-a-tool-for-human-conquest-of-nature world view that had dominated political thought for centuries. On the other hand, it provided, almost without anyone realizing it, a transcendent but almost-subversive image of Earth – breathtakingly beautiful, but also extraordinarily fragile – an image that steadily seeped into the collective consciousness of people everywhere. It fundamentally re-shaped the common understanding of how interdependent life on Earth was, and of how tenuous was our existence on this our only planet. Neil Armstrong, Commander of the Apollo 11 flight, and first human to walk on the Moon.

> It suddenly struck me that that tiny pea, pretty and blue, was the Earth. I put up my thumb and shut one eye, and my thumb blotted out the planet Earth. I didn't feel like a giant. I felt very, very small.
>
> (Armstrong, 1969)

As the new decade of the 1970s began, millions of people gathered across the United States for the first Earth Day (22 April 1970), organized by Gaylord Nelson, a senator from Wisconsin, and Denis Hayes, a Harvard graduate student.

The following year UN Secretary-General U Thant supported the global initiative to celebrate this annual event, signing a proclamation saying:

> May there be only peaceful and cheerful Earth Days to come for our beautiful Spaceship Earth as it continues to spin and circle in frigid space with its warm and fragile cargo of animate life.
>
> (Thant, 1970)

By 1972, three of what would become the major international environmental NGOs were formed – Greenpeace, Friends of the Earth, and the International Institute of Environment and Development.

The preparation

The stage for the Environmental Conference in Stockholm was set by three extraordinary reports. The first was *Limits to Growth* (Meadows *et al.*, 1972) published by the Club of Rome. It tracked five variables – world population, industrialization, pollution, food production, and resource depletion – and found the trends for each of them leading rapidly to a point of environmental, economic or social collapse. The report was perhaps ahead of its time and caused widespread controversy.

Thirty five years later, however, the Commonwealth Scientific and Industrial Research Organization in Australia examined the 1972 prediction of the Club of Rome report, and found that they were all consistent with the world's path through the first decade of the twenty-first century (Turner, 2008).

The second influential report was commissioned by Maurice Strong, Secretary-General of the approaching Conference in Stockholm. Written by Barbara Ward (who had co-founded the International Institute of Environment and Development) and René Dubos (who authored the phrase 'think globally and act locally'), it argued that while the degradation of the environment in industrialized countries derived from production and consumption patterns, the environmental problems in the rest of the world were largely the result of underdevelopment and poverty. It called for the integration of development and environmental strategies, and argued to rich nations that it was in their own interests to provide additional funding to help enable poorer nations to achieve sufficient economic growth to allow their people to focus on conserving their natural resources.

It was called *Only One Earth* and it sounded an impassioned call to arms, an urgent alarm about the impact of human activity on the planet. The authors, however, maintained a tone of optimism that humans could together address the challenges we faced. The slogan, 'Only One Earth', also became the rallying cry of Stockholm, and – as relevant today as it was then – it is the title we take for this book.

The final report that helped shape the 1972 Conference was produced from an in-depth seminar held Founex, Switzerland. Built on commissioned papers by experts from within and outside the UN system, it addressed the themes of the Stockholm Conference, and it played a significant role in persuading developing countries to engage, by making a compelling case for the relationship between environment and development and the need for environmental policy to contribute to the sustainable development of developing countries. One of the report's most significant sentences was:

> If the concern for human environment reinforces the commitment to development, it must also reinforce the commitment to international aid.
>
> Founex Report (Stoddart, 1971)

Gaining serious, high-level participation by governments was the Stockholm Conference's first challenge. Developed countries like France, Germany, Canada, the Nordics, the UK and the United States – where the environmental movement had already gained public support – were on board from the start. However, it turned out that they had intended to ensure that the conference did not go too far in giving rise to demands for significant support from more developed countries, by convening a group known as the Brussels Group. This group it seems met as 'an unofficial policy-making body to concert the views of the principal governments concerned', according to a note of one of the group's first meetings written by a civil servant in the British Foreign and Commonwealth Office. 'It will have to remain informal and confidential' (Hammer, 2002). This meeting took place in July 1971, nearly a year before the Stockholm conference opened.

Developing countries were far more cautious about participating. Brazil concluded it needed to attend, if only to defend its natural resources against international regulation. Participation by sceptical developing countries was strongly influenced by the decision of India's Prime Minister Indira Ghandi to attend in response to persuasion by the conference Secretary-General on the grounds that she would bring to the conference the most influential voice of the developing world.

The conference

A dispute on the political status of the German Democratic Republic led the Soviet Union and the eastern European governments to decide they would not participate. The US and the UK had contributed to a boycott by the Soviet bloc. In the previous UN General Assembly they invoked the 26-year-old Vienna Formula for who should attend the conference. This required countries to be a member of the UN or its specialized agencies. West Germany was already a member of UNESCO and WHO, so could attend; but East Germany was not. This resulted in the boycott by the Soviet Union and the Eastern Bloc countries. Both East and West Germany were to join the UN the following year on 18 September 1973.

On the other hand, the People's Republic of China did attend – its first major world conference since taking its seat at the United Nations.

Strong was fortunate in his choice of senior staff and advisors. His chief of staff, Marc Nerfin, a young non-establishment Swiss, brought a combination of sound judgement, radical yet pragmatic social views which earned the support of important constituencies and the respect of all. He helped to organize and enlist the participation of a number of key developing country's experts, most of whom had been critical of the conference as focusing on an issue of principal interest to the more developed countries, which could divert attention and resources from the development and relief of poverty which were the priorities of developing countries. These included Professor Ignacy Sachs of the Sorbonne, Ambassadors Almeida Ozorio and Bernardo Ritto of Brazil, Gamani Correa of Sri Lanka, and Mahboub ul Haq of Pakistan, and from the more developed countries the young Australian James Wolfensohn, who later became President of the World Bank. The distinguished British economist Lady Jackson (Barbara Ward) invoked her star quality influence in helping to bring the group together and lead its deliberations, which were focused on how best to engage them constructively in the conference and its preparations.

The political challenge of the Conference was whether developed countries could accommodate the interests of the developing countries.

Two countries were key to the success of the conference. Brazil had taken a strong sceptical position, derived from its well-reasoned policy analysis that the conference might attempt to impose constraints on its industrial resources and agricultural development. Brazil was seen as the 'custodian' of the Amazon basin – an extraordinary global resource. Along with other developing countries, Brazil had argued that there was a fundamental right for developing countries to exploit their own natural resources, just as developed countries had. If there was international requirement for environmental protection, Brazil and others maintained that developed countries were obligated to provide developing countries with additional financing and access to the technologies required to evaluate them do this.

This framed what appeared to be an intractable conflict between environmental conservation and economic development, between global needs and national rights, between wealthy and poor nations.

Addressing this conflict – and simultaneously pointing a way to its resolution – Prime Minister Gandhi of India provided the link that was vital to establish the connection between poverty and the environment. In her opening statement she made one of the most influential presentations of the entire conference, on the theme that 'poverty is the greatest polluter of all'.

This set in motion two weeks of negotiations that moved towards recognition that national governments had obligations – and jurisdiction – over environmental policy within their borders, but that the international community was responsible for supporting developing countries in carrying out those obligations.

Not all the criticisms of Stockholm came from developing countries. Prominent amongst critics and detractors from the more developed countries was the influential and very vocal economist Barry Commoner. On the other hand it proved challenging to overcome the widespread apathy and scepticism about the conference and its prospects. Despite this the conference itself received significant publicity in part due to North–South conflicts which threatened the prospects of agreement and the wide differences amongst so many participants which became evident. The conference newspapers summed this up well in its heading after the first day of the conference 'Only 113 Earths'. As the conference progressed, however, these differences were both highlighted and accommodated through an extensive round of negotiations both in the committees of the conference and in informal consultations.

This was the first major UN conference in which non-governmental organizations (NGOs) – as stakeholders – had a significant influence. They were concentrated in an area called the 'Hog Farm', but the Secretariat made every effort to facilitate their interaction with delegates and their contributions to the negotiating process. Strong made a point of visiting the farm and addressing them there. He also enabled Barbara Ward as their most influential spokesperson to address the delegates.

The Chinese tried to address a roadblock over the Stockholm Declaration by placing a 'non paper' on the table – their first at a UN meeting.

Box 1.1 **The Chinese declaration**

1 Relationship between economic development and environment

Economic development and social progress are necessary for the welfare of mankind and the further improvement of the environment. The developing countries want to build a modern industry and agriculture to safeguard their national independence and assure their development. A distinction must be made between these countries and the more highly developed countries. The environmental policies of each nation must not impede development.

2 Population growth and environmental protection

Man is the most precious of all things on earth. Man propels social progress, creates social wealth and advances science and technology. There are no grounds for any pessimistic views on population growth and the preservation of the environment. Both can be solved by national policies: the control of urban population, settlement in agricultural areas, publicity for the environment, avocation of family planning, etc.

3 Social root cause of environmental pollution

We hold that the major root cause of environmental pollution is capitalism, which has developed into a state of imperialism, monopoly, colonialism, and neocolonialism – seeking high profits, not concerned with the life or death of people, and discharging poison at will. It is the policies of the superpowers that have resulted in the most serious harm to the environment. The United States has committed serious abuses in Vietnam, killing and wounding many inhabitants. These facts are known to the world and should be included in the Declaration. The Declaration should also be comprehensive on the nuclear threat.

4 Protection of resources

Every country should be entitled to utilize and to exploit its resources for its own needs. We resolutely oppose the plundering of resources in the developing countries by the highly developed countries.

5 Struggle against pollution

The governments of all countries must take steps to prevent the discharge of pollutants into the environment. We support the people of all countries who are struggling against pollution.

continued ...

Box 1.1 continued

6 International pollution compensation

Each country has the right to safeguard its environment. The corporate states are discharging pollutants and the victim states have a right to compensation.

7 International exchange of science and technology

All countries should actively employ science and technology to safeguard the environment. Science and technology should not be monopolized by one or two countries. They must be available for the protection of the environment in the developing countries.

8 Voluntary fund

The creation and management of an environmental fund has been proposed. The principal industrialized countries and the most serious polluters should make the contribution to this fund.

9 World environment body

The intergovernmental body which has been proposed to guide and direct environmental policy must respect the sovereignty of all countries. This body must be free from the control by the super powers.

10 International environmental protection and state sovereignty

Any international agreement should respect the sovereignty of all countries. No country should encroach on another under the pretext of environmental protection.

On the last day of the meeting, representatives of the Chinese delegation announced that they could not take a final position on the declaration because of its reference to population. They didn't have time to get authorization from Beijing because of the time difference between there and Stockholm. They saw no alternative but to walk out of the final session.

This would have left a dark cloud hanging over the proceedings. Maurice Strong suggested a way to finesse the problem:

> When the final vote is called, your entire delegation could rise from its seats at the table, but instead of leaving altogether you'd merely move to the seats immediately behind you [seats normally occupied by non-voting staff].

You'd not register a vote, but you wouldn't abstain either, or walk out of the meeting. The records of the meeting could then show you as present and implicitly, but not explicitly, participating in the consensus. For your purposes, on the other hand, it couldn't be shown that you had voted in the absence of authorization from your government.

(Strong, 2001)

It was agreed to. In the end, the Stockholm Conference was attended by the representatives of 113 countries, 19 inter-governmental agencies, and more than 400 inter-governmental and non-governmental organizations.

It is widely recognized as the beginning of modern political and public awareness of global environmental problems, and placed them firmly on the international agenda.

Outcomes

The conference produced three major outcomes – a Declaration and Statement of Principles, an Action Plan, and a new United Nations Programme on the Environment.

The Declaration consisted of 27 principles. It declared that:

The United Nations Conference on the Human Environment ... [had] considered the need for a common outlook and for common principles to inspire and guide the peoples of the world in the preservation and enhancement of the human environment...

(United Nations, 1972)

Its major conclusion was that

Local and national governments will bear the greatest burden for large-scale environmental policy and action within their jurisdictions. International cooperation is also needed in order to raise resources to support the developing countries in carrying out their responsibilities in this field. A growing class of environmental problems, because they are regional or global in extent or because they affect the common international realm, will require extensive cooperation among nations and action by international organizations in the common interest.

(United Nations, 1972)

The Declaration was approved by consensus.

What many sceptics assumed would only be a rhetorical statement had emerged as a highly significant statement of principles, policies and goals. They reflected a new community of interest among nations, regardless of their politics, ideologies or economic status. Despite the difficulties and the differences that had emerged in achieving them – the very fact that delegates laboured as they did – testified to the importance their governments attached to the Declaration and to the very basic recognition that it is every nation's responsibility to ensure that activities within its jurisdiction or control do not cause damage to the environment of other states or of areas beyond the limits of national boundaries.

The Action Plan agreed by the conference included 109 recommendations. Among them, it set into motion mechanisms that eventually would:

- drastically curtail emissions into the atmosphere of chlorinated hydrocarbons (CFCs) that destroy the ozone layer, and heavy metals;
- provide information about possible harmful effects of various activities before these activities are initiated;
- accelerate research to assess the risk of 'climate modification' and begin consultations among concerned countries;
- assist developing countries to cope with urban priorities needs such as housing, water supply and waste disposal;
- intensify the preparation of conventions on the protection of the world's natural and cultural heritage;
- prioritise education and information to enable people to weigh the decisions which shape their future and to create a wider sense of responsibility;
- initiate steps to protect and manage common resources, considered of unique value to the world community;
- initiate a global programme to ensure genetic resources for future generations;
- create an International Referral Service for nations to exchange environmental information and knowledge;
- incorporate environmental considerations into the review of the development strategies embodied in the Second Development Decade;
- pursue regional cooperation for financial and technical assistance;
- prevent environmental standards from becoming pretexts to limit trade or impose barriers against exports of developing countries;
- study the financing of additional costs to developing countries arising from environmental considerations.

The conference also led to the establishment of the first UN body to deal solely with the environment. The United Nations Environment Programme (UNEP) was given the mandate to coordinate the development of environmental policy consensus by keeping the global environment under review and bringing emerging issues to the attention of governments and the international community for action. Governments were determined to keep the UNEP small and limit both its funding and staffing. It was initially set up with a fund of US$100 million for the first five years, less than the US$164 million asked for by Maurice Strong. Comparing it with the budget for the US Environmental Protection Agency (EPA) for the same period is interesting – its budget was US$64 billion. In fact a representative of the US State Department admitted that the US$20 million a year will be 'inadequate to launch all appropriate programmes simultaneously' (ECO, 1972a). The decision to establish its headquarters in Nairobi far from the various organizations with which it had to inter-act and influence also constrained it.

Box 1.2 **Whales reported killed 1969/1970**

USSR	18,336
Japan	17,047
Peru	1,935
South Africa	1,880
Australia	799
Canada	712
Spain	413
Iceland	337
Portugal	249
Norway	228
Chile	103
Brazil	102
US	73
Total	42,254

(ECO, 1972)

Even with funding problems, from the beginning UNEP has played a significant role in developing international environmental regulation and environment awareness in areas such as climate change, ozone depletion, oceans, biodiversity and toxic chemicals. The creation of the Environment Coordination Board included the leaders of the specialized agencies to influence and facilitate coordination of their programmes and politics. They were resistant to this, but the fact that both the President of the World Bank and the Managing Director of the International Monetary Fund agreed to participate in person made it essential for other agency heads to do so.

The European Union followed the UN's model in 1973, with the establishment of a directorate on the environment. National governments also started to set up environmental ministries, with the first being the UK.

Stockholm called for a Convention on the Prevention of Marine Pollution by Dumping of Wastes and Other Matter. By the end of 1972 a draft was completed, and the convention came into force in 1975 when 15 countries had ratified it.

The Conference also called for a moratorium on whaling. The International Whaling Commission was scheduled to meet immediately after Stockholm, and conference Secretary-General Maurice Strong travelled and addressed the Commission, uninvited, and succeeded in getting it to agree to a limited moratorium.

A full (if not permanent) moratorium came in 1986. By 2010 less than 1000 were being killed for 'scientific research'.

Stakeholders

As important as all of these, were the advances for what we have come to know as civil society or stakeholders. In the run up to the conference, members had played a significant role in the Founex Report as well as mobilizing public opinion on environmental issues. The UN Secretariat included as special advisor for NGOs, Henrik Beer, the Secretary-General of the International Federation of Red Cross Societies, as well as Baron von dem Bussche, a distinguished German who volunteered his effective and influential services.

To support the work of NGOs at the conference, a space was provided by the Swedish government which became the 'Environment Forum'. Stockholm marked the first real entry of environmentalists into the international policy arena. Their ability to galvanize public opinion earned them a continuing role in influencing the decision-making process thereafter. Their achievements in Stockholm also gave rise to national environmental movements in countries which prior to the Stockholm Conference had not had any. At Stockholm, many environmentalists had argued for a 'steady state' theory of economics, a 'no growth strategy'. Interestingly, in the preparations for the Earth Summit in 2012, many NGOs and economists have returned to that discussion.

Conclusion

It can be said that the Stockholm Conference was the birth of the environment movement worldwide, whether it's Greenpeace, Friends of the Earth, Earth Day, UNEP, US EPA and other EPAs, the creation of environment ministers in government, and environmental journalism; it all started around the same time as the conference.

In the last NGO *ECO* newsletter the NGOs said the conference should be judged by four questions:

1   Can populations go on growing indefinitely?
2   Is there an infinite supply of non-renewable energy?
3   Are ecosystems infinitely flexible, infinitely resilient?
4   Does the socio-economic systems they support provide optimum satisfaction for all its members?

The answer to all four is, of course, no – yet invariably politician decisions rest on the assumption that it is yes

If we accept ecological reality, we must also accept a fundamental reform of society and economic system which drives it. If everyday decisions begin to be taken in light of these conditions, this conference will be partly responsible. And that's quite an achievement.

(ECO, 1972)

Chapter 2

The road to Rio

Man is both creature and molder of his environment, which gives him physical sustenance and affords him the opportunity for intellectual, moral, social and spiritual growth. In the long and tortuous evolution of the human race on this planet a stage has been reached when, through the rapid acceleration of science and technology, man has acquired the power to transform his environment in countless ways and on an unprecedented scale. Both aspects of man's environment, the natural and the man-made, are essential to his well-being and to the enjoyment of basic human rights the right to life itself.

Stockholm Declaration (United Nations, 1972)

The Stockholm Declaration was a radically far-sighted document for its era, and at the time of its agreement few people fully grasped the sweeping implications of its principles. It opened the way to an active debate on the difficult choices facing societies about the ecological imbalances created by human activities, about equity in the use of common resources, and about the sharing of power between countries and within them. It dared to address the fundamental purposes of economic growth on this, the one planet all people live on and how peoples, species and nations might share the benefits and the costs – decisions that all still remain to be made. It also started to ask fundamental questions about who should participate in the decision making.

Public awareness of the environment had grown considerably since Sweden tabled the resolution for a UN Conference on the Environment in 1968. Before Stockholm, virtually no newspaper had designated an environment correspondent, virtually no company was producing environmental reports, and virtually no environmental courses were being taught in schools. Stockholm changed all that.

Creation of UNEP

Although the Stockholm Conference called for the establishment of a United Nations Environment Programme, it was the UN General Assembly which formally established it legally. On 15 December 1972 the UN General Assembly passed resolution 2997 (XXVII) setting out the mandate and objectives of the new body.

The organizational capacity of the new programme was kept weak. A group of European countries which supported its establishment, including Britain, the US, Germany, Italy, Belgium, the Netherlands and France, had agreed secretly to ensure that it would not have the support required.

The group was concerned that any new environmental regulations would have an impact on trade. They also wanted to ensure that UNEP did not have a large budget as it would then be restricted on what it could do.

In a paper written for the Brussels Group in September 1971, the UK government said that a

> new and expensive international organization must be avoided, but a small effective central coordinating mechanism ... would not be welcome but is probably inevitable.
>
> (Hamer, 2002)

The new UNEP was empowered to investigate problems and recommend strategies to solve them, but it was not authorized to implement those strategies on the ground. While other UN bodies – such as the United Nations Development Programme (UNDP) – were set up precisely for project design and implementation, UNEP was expressly limited from providing services. Although UNEP carried the word 'programme' in its name, it was not established as a UN programme-level agency.

Even though a member of the Brussels Group, perhaps under-appreciated, is the leadership role the United States, under President Richard Nixon, had played in the follow-up to Stockholm. The US wanted to host UNEP in Washington; this would have had a profound impact on the ability of the organisation to interface with the World Bank and UNDP. The US wasn't the only country wanting to host the new UN programme. India and Kenya also offered to host it

But in international policy, 1972 was a time of transition. The era of colonialism had just ended with the establishment of newly independent nations in Africa, Asia and the Pacific in the 1950s and 1960s. There was intense competition between the US and the Soviet Union to win diplomatic support from those nations. For those countries – and for the emerging political bloc of the Third World, or 'non-aligned' movement (NAM) – it had become a point of contention that all of the major UN agencies were based in Northern-hemisphere, developed nations.

The previous year the United Nations Industrial Development Organisation (UNIDO) was created and was established in Vienna. So it was widely agreed that the new organization be headquartered in a developing country. The two developing countries that sought to host were India and Kenya.

Despite the consensus, many developed countries were worried about the choice of Kenya. It was distant from other intergovernmental organizations that it needed to work with, and the telecommunications system into and within Nairobi was notoriously inefficient. Other concerns included the ability to attract qualified staff and basic questions of personal security. These were complex matters that needed to be addressed, but even mentioning such issues risked touching on deep regional sensitivities.

Partly because of his reputation for solving logistical and political challenges, Maurice Strong was approached to be the new organization's first Executive Director. Prime Minister Pierre Elliott Trudeau of Canada agreed to let Strong be the Executive Director for one term to get UNEP running, as opposed to returning to head the Canadian International Development Agency (CIDA).

The General Assembly unanimously approved Strong as UNEP's first Executive Director. Mostafa Tolba, a biologist who had led Egypt's delegation to Stockholm, became his deputy and Philip Ndegwa, a talented Kenyan economist, was made Senior Policy Advisor. The new organization moved to deal with the logistics of its new UN home. Starting with a contribution of US$100 million from governments, it established initially in the centre of Nairobi while its new home was being built in Gigiri just outside Nairobi.

UNEP took an early initiative to establish an outer limits programme which examined global risks that needed to be addressed. One of the first was climate change. In 1973 UNEP brought together a meeting of scientists to review the evidence on the issue – this gave UNEP an early leadership role on the issue.

In 1975, Maurice Strong resigned, in accordance with the agreement made with his government, to become CEO of Petro-Canada. Mostafa Tolba was appointed to succeed him. He served until 1992, playing a critical leadership role in creating a landscape of multilateral environmental agreements.

Oil crisis

In what would seem to become an inverse symmetry after each UN environment and sustainable development conference, the year after Stockholm brought a stock market crash and oil crisis. It was as if the environment picked the worst times to engage the priority of governments. In January 1973, a stock market crash was triggered by inflation and the withdrawal of the US from the Bretton Woods Accord under which the US dollar had been pegged to the gold standard. On 6 October, a surprise attack by Syria and Egypt on Israel launched the Yom Kippur War. Ten days later, the Organization of Petroleum Exporting Countries (OPEC) announced that they were going to raise oil prices by 70 per cent, would reduce production by 5 per cent, and would continue to cut production until they achieved their economic objective of a higher return for oil, and their political objective of forcing a return to pre-1967 borders in the Middle East.

The oil embargo against Western countries continued for 17 months, and the resulting recession continued in many countries until the beginning of the 1980s. The impact this had in distracting public attention from environmental issues and diverting political will from follow-up to the Stockholm Conference agreements cannot be over-stated. The political and economic impacts of the oil crisis were profound. The 'Western' bloc of industrialized nations shifted from an era of almost continuous growth and prosperity to a period of economic uncertainty and increasing political turmoil. The economic pressure and the perceived threat to national security energized the rise of conservative political leaders who called for reductions in social spending, lowering taxes for corporations and the rich, and increases in military investment.

By the end of the decade, Ronald Reagan had been elected President of the United States and Margaret Thatcher had become Prime Minister of the UK on platforms based on radical 'liberalization' of their markets and de-regulation of their social protections in order to make their economies and militaries 'stronger'. In such a context, there was little political interest or will to invest in promoting new international institutions to help development in poorer countries, or to protect the environment.

The idea that increasing investment in energy sources other than petroleum could effectively address threats to energy security received only fleeting consideration. The possibility that improving the lives of millions in impoverished countries could reduce military threats was ignored. The counter-argument of nationalist politicians that environmental safeguards restricted the growth of business, and that the need to support domestic economies obviated any interest in international development or cooperation predicted all too clearly the even more extreme claims of powerful factions in those and many other countries today.

International environmental governance

One of the major achievements from 1972 to 1992 was the negotiation of an extraordinary number of multilateral environmental agreements (MEAs). These conventions generally divided into six thematic clusters: oceans and regional seas, freshwater, biodiversity, atmosphere, land, and chemicals and hazardous wastes.

The MEA processes adopted after 1972 generally each contain the following key elements:

- a Conference of the Parties (COP);
- a secretariat;
- one or more advisory bodies;
- a clearing-house mechanism; and
- a financial mechanism.

Some of the significant MEAS developed from 1972 to 1992 are:

- marine environment and regional seas
  - United Nations Convention on the Law of the Sea (UNCLOS), 1982;
  - Convention for the Protection of the Marine Environment of the North-east Atlantic OSPAR, Paris, 1992;
  - Convention on the Protection of the Marine Environment of the Baltic Sea Area 1992;
  - Helsinki Convention, Helsinki, 1992;

- Conventions within the UNEP Regional Seas Programme;
- Convention on the Protection of the Black Sea against Pollution, Bucharest, 1992;
- Convention for the Protection and Development of the Marine Environment of the Wider Caribbean Region, Cartagena de Indias, 1983;
- Convention of the Protection, Management and Development of the Marine and Coastal Environment of the Eastern African Region, Nairobi, 1985;
- Convention for the Protection and Development of the Marine Environment and Coastal Region of the Mediterranean Sea Barcelona Convention, Barcelona, 1976;
- Convention for the Protection of the Natural Resources and Environment of the South Pacific Region, Nouméa, 1986;
- Convention for the Protection of the Marine Environment and Coastal Area of the South-east Pacific, Lima, 1981;
- Convention for Co-operation in the Protection and Development of the Marine and Coastal Environment of the West and Central African Region, Abidjan, 1981;
- Framework Convention for the Protection of the Marine Environment of the Caspian Sea;
- Kuwait Regional Convention for Co-operation on the Protection of the Marine Environment from Pollution, Kuwait, 1978;
- Regional Convention for the Conservation of the Red Sea and the Gulf of Aden Environment, Jeddah, 1982; and
- Convention for the Conservation of Antarctic Marine Living Resources (CCAMLR), Canberra, 1980.
- freshwater
  - Convention on the Protection and Use of Transboundary Watercourses and International Lakes (ECE Water Convention), Helsinki, 1992.
- biodiversity
  - Convention on International Trade in Endangered Species of Wild Flora and Fauna (CITES), 1973;
  - Convention on the Conservation of Migratory Species of Wild Animals, (CMS), Bonn, 1979;
  - Convention to Combat Desertification (CCD), Paris, 1994; and
  - International Tropical Timber Agreement, (ITTA), Geneva, 1994.
- atmosphere
  - Vienna Convention for the Protection of the Ozone Layer, 1985;
  - Montreal Protocol on substances that deplete the Ozone, 1987;
  - Amendment to the Montreal Protocol (London), 1990; and
  - Amendment to the Montreal Protocol (Copenhagen), 1992.
- chemicals and hazardous wastes
  - Basel Convention on the Control of Transboundary Movement of Hazardous Wastes, 1989;
  - Rotterdam Convention, 1988;

   - Bamako Convention on the Ban of the import into Africa and the control
     of transboundary movement and management of hazardous wastes within
     Africa, 1991.

Although the coherence of the MEA issue 'clusters' seems obvious in retrospect,
there was little attempt to group the conventions together to deal with them more
efficiently. Siting the commonly themed convention secretariats in the same venue
could have resulted in both logistical efficiencies for participants and increased
synergies from close communications among secretariat staff. As UNEP played a
significant role in advocating and developing many of the MEAs, their secretariats
might have all been housed in Nairobi, or a few other central locations. Besides the
efficiencies of cooperation and scale, this would have strengthened the role of UNEP.

What emerged instead was precisely the opposite. As each MEA was negotiated, a
number of countries competed to host them. The result was a fragmented system of
MEAs where conventions dealing with similar issues were each located in different
cities and countries – and a squandered opportunity to increase UNEP's ability to
project a strong, focused voice for the environment.

Perhaps the most significant of the MEAs of this decade as far as its substantive
and political achievement was the 1985 Vienna Convention for the Protection of
the Ozone Layer, and its 1987 Montreal Protocol on Substances that Deplete the
Ozone Layer.

The history of the ozone issue is well known. For years researchers had warned
that the steady release into the atmosphere of chlorofluorocarbons (CFCs) used in
the manufacture and use of plastics, refrigerants and aerosol spray products could be
reacting with and destroying ozone molecules high in the earth's atmosphere. These
molecules, while highly dispersed, formed a critical ozone layer that absorbed large
percentages of dangerous ultraviolet solar radiation and blocked it from reaching the
planet's surface.

As would virtually be paralleled in the later debate on climate change, powerful
chemical industry groups actively opposed action on CFCs, fearing a loss of sales
and profits. They argued that: a) the science was unproven, b) a ban would prove
economically devastating, and c) discovering replacement technologies would take
decades. The chairman of the US-based chemical multinational DuPont claimed
that ozone depletion theory was 'a science fiction tale … a load of rubbish, utter
nonsense' (Roan, 1989: 56). Governments – in particular the US – had responded
to industry's complaints by blocking progress in negotiations.

In 1985, the British Antarctic Survey published results that revealed a growing
'ozone hole' that seasonally expanded and contracted in the stratosphere above the
South Pole, and had begun to reach populated areas. Later that year, 20 governments
– closely tracked by major industry CFC producers – met in Vienna and negotiated
a framework convention, motivated in part by reaction to the lengthy deliberations
for the United Nations Convention on the Law of Sea (1972–1983). The framework
process allowed for individual protocols with specific targets to be added as and
when the international community was ready.

In 1986, the Alliance for Responsible CFC Policy (an association founded by DuPont to represent the CFC industry) argued that the science was too uncertain to justify any action. In 1987, DuPont testified before the US Congress that 'we believe that there is no immediate crisis that demands unilateral regulation' (Wikipedia, Montreal Protocol).

Today the Alliance describes itself as:

> Overall the Alliance has advocated the benefits of alternatives to CFCs; educated policymakers as to the feasibility of laws and regulations, assisted in removing barriers to the use of many alternatives; focused the US government on curtailing illegal trade in CFCs; guided government efforts on many regulatory issues, supported an annual conference for discussion on policy and technology developments, and disseminated information to the public, media, governments, and industry that is useful in making the transition to safer alternatives.
>
> (Alliance for Responsible CFC Policy, n.d.)

A change of position happened in 1988. After Du Pont began producing hydrofluorocarbons (HCFCs) as profitable alternatives to Freon – and as new scientific evidence emerged that ozone depletion was worse than reported – the company suddenly announced a phase-out of CFCs. The US, under President Ronald Reagan, quickly decided that a specific ban was desirable. And within a year the Montreal Protocol was negotiated with effective phase-outs of the most damaging CFCs.

> The history of the Montreal Protocol demonstrates how the international community can unite to successfully solve critical environmental problems. We must learn from our past successes, tackle current and future challenges, and remember that what is at stake in the battle against climate is our existence.
>
> (Picolotti quoted in Kaniaru, 2007: 358)

It also demonstrates how the actions of non-governmental campaigners and the scientific community are absolutely critical to pushing governments – and finally convincing powerful business interests – to accept their responsibilities to acknowledge dangerous problems and to act.

Economic development

The 1970s recession was confronted by developing countries' call for a new financial deal on trade, debt and aid.

Developed through the United Nations Conference on Trade and Development (UNCTAD), there was an attempt to restructure North–South economic relations. This resulted in a call for a new international economic order (NIEO). The term was derived from the *Declaration for the Establishment of a New International Economic Order*, adopted by the United Nations General Assembly in 1974, and referred to a wide range of trade, financial, commodity and debt-related issues (United Nations, 1974).

> The present international economic order is in direct conflict with current developments in international political and economic relations. Since 1970 the world economy has experienced a series of grave crises which have had severe repercussions, especially on the developing countries because of their generally greater vulnerability to external economic impulses. The developing world has become a powerful factor that makes its influence felt in all fields of international activity. These irreversible changes in the relationship of forces in the world necessitate the active, full and equal participation of the developing countries in the formulation and application of all decisions that concern the international community.
>
> (United Nations, 1974)

The objective of NIEO was to improve the terms of trade for developing countries; to see an increase in the overseas development assistance to the promised 0.7 per cent of gross domestic product (GDP) and a reduction in tariffs by developed countries.

One of the principal calls of the Declaration was for ensuring the ability of developing countries to control the activities of multinational companies operating in their borders. Among the few outcomes from NIEO was the development of the Restrictive Business Practice Code, adopted in 1980. Another dealt with commodities. In 1980, The Agreement Establishing the Common Fund for Commodities (CFC) was adopted (27 June 1980) in Geneva by the United Nations Negotiating Conference on a Common Fund. It was set up to increase cooperation on commodities, but took until 1989 to come into force.

Meanwhile, green economic theory was beginning to develop in the years after the publications of the Club of Rome's *Limits to Growth* (Meadows *et al.*, 1972) and E.F. Schumacher's *Small Is Beautiful: A Study of Economics as if People Mattered* (1973). These were followed in 1977 by Herman Daly's comprehensive and persuasive *Steady-State Economics* (1977).

Ten years from Stockholm

To review progress, to celebrate Stockholm's achievements and to plan an agenda for the next decade, UNEP hosted the 10th Session of its Governing Council in 1982. The ten years following the Stockholm conference had been marked by increased general acceptance of the once-revolutionary concept of environment-based development. Moreover, the commitment of international and national development assistance institutions directly funding only sustainable projects was noted as 'landmark'.

One of UNEP's major achievements during this period was the progressive establishment of a centre for environmental information, which was accomplished vis-à-vis its environmental assessment and monitoring programmes.

UNEP had also taken the lead in a series of useful initiatives, addressing rapidly emerging global issue areas, including the atmosphere, and most notably the build-up of carbon dioxide in the atmosphere and the depletion of the ozone layer.

It was also suggested that the next environmental decade, or the 'Nairobi Decade', should highlight the environmental needs and concerns of developing countries, as well as establish more regional and sub-regional programmes. The Nairobi Declaration framed the challenge:

> During the last decade, new perceptions have emerged: the need for environmental management and assessment, the intricately complex interrelationship between environment, development, population and resources and the strain on the environment generated, particularly in urban areas, by increasing population have become widely recognized. A comprehensive and regionally integrated approach that emphasizes this interrelationship can lead to environmentally sound and sustainable socio-economic development.
>
> Nairobi Declaration (UNEP, 1982)

The basic focus for the next decade, drafted by the UNEP GC's 10th session, encouraged the promotion of policies and programmes in information, education, training, and national institution-building, and coordinating such policies for rational resource and environmental management as an 'integral' part of social and economic development. Such strategies applied to the area of environmental assessment were expected to improve 'early warning indicators of significant environmental changes, the planning and co-ordination of monitoring at the global and regional levels, and produce concrete assessment statements for environmental topics and their human health'. The Session similarly proposed establishment of stronger links between the Global Environmental Monitoring System, the International Referral System for sources of environmental information, the International Register of Potentially Toxic Chemicals and national and international data centres.

But probably the most consequential recommendation of the 1982 UNEP Special Session was a suggestion tabled by the Canadian Government that a special commission should be set up to look at:

> long term environmental strategies for achieving sustainable development to the year 2000 and beyond.
>
> (UNEP, 1982a)

This became the first UN document to mention sustainable development. But its substance was even more significant.

In 1983, the UN General Assembly established the World Commission on Environment and Development (WCED) to be chaired by Norwegian physician and former Prime Minister, Gro Harlem Brundtland (United Nations, 1983).

Its terms of reference suggested that the Special Commission should:

- propose long-term environmental strategies for achieving sustainable development to the year 2000 and beyond;
- recommend ways in which concern for the environment may be translated into greater co-operation among developing countries and between developing and developed countries, and lead to objectives which take account of the interrelationships between people, resources, environment and development;
- consider ways the international community can deal more effectively with environmental concerns; and
- help define long-term environmental issues and a long-term agenda for action during the coming decades, and aspiration goals for the world community.

(United Nations, 1982: 8. (a)–(d))

The ground-breaking report the Commission produced, *Our Common Future* (WCED, 1987), became known as the Brundtland Report, and would provide the conceptual and political framework for integrating a vast panoply of ecological, social, economic, participation, governance, and even lifestyle issues – and for changing the way governments and average individuals looked at their planet and its possibilities for its future development.

Our Common Future

The term 'sustainable development' first came to prominence in 1980, when the International Union for the Conservation of Nature and Natural Resources (IUCN) agreed its World Conservation Strategy. This established 'the overall aim of achieving sustainable development through the conservation of living resources' (IUCN, 1980):

The 'environment' is where we live; and development is what we all do in attempting to improve our lot within that abode. The two are inseperable.

Our Common Future (WCED, 1987)

The World Commission on Environment and Development was structured to include prominent individuals from NGOs, business, academia and governments around the world. Its Chair was former Norwegian Prime Minister Gro Harlem Brundtland and its Vice Chair was Mansour Khalid, the Deputy Prime Minister of Sudan. Of particular importance was the role of its Secretary-General, the highly experienced and competent Canadian James MacNeill. Parallel to the Commission were three advisory panels – on energy, industry and food security – to assist with their deliberations.

The Commission was the first to conduct hearings that extensively engaged civil society. In total there were fifteen public hearings held across the world to ensure that the input was from society as a whole. Running these was Chip Lindner, secretary to the Commission and former Deputy Director of WWF International. After the Commission Report was published, the Centre for Our Common Future (COCF) was set up to continue to work indentified by the Commission. It became the obvious body then to help engage stakeholders in the Rio Earth Summit.

Our Common Future called for a new form of development – sustainable development – and described a change of policies that would be required for achieving that. Its definition became one of the most widely used over the coming years:

> Sustainable development is development that meets the needs of the present without compromising the ability of future generations to meet their own needs. It contains within it two key concepts:
> * the concept of 'needs', in particular the essential needs of the world's poor, to which overriding priority should be given; and
> * the idea of limitations imposed by the state of technology and social organization on the environment's ability to meet present and future needs.
> Our Common Future (WCED, 1987)

The Commission's most influential recommendation was to call for an international convention on environmental protection and sustainable development.

In December 1989, the General Assembly formally agreed to convene another global conference, formally titled the United Nations Conference on Environment and Development (UNCED), but soon to be universally known as the Earth Summit. Even before the General Assembly met, the Canadian government proposed that Maurice Strong serve as Secretary-General of the conference in 1992 – the same capacity he had held in 1972, as Secretary-General of the Stockholm Conference. He was appointed Conference Secretary-General in February 1990. Strong proposed as chair of the conference's preparatory committee (PrepCom) Tommy Koh of Singapore, known for his masterful chairing of the third Law of the Sea conference, in which he had brokered deals bridging substantial disagreements between North and South, East and West, and land-locked and coastal states. Most of the work for the conference was conducted by the Secretariat and the PrepCom, which held four extraordinary and extensive sessions from August 1990 to April 1992.

The extraordinary substantive scope and the political complexity of the UNCED agenda made it a virtually unprecedented challenge. Strong knew from experience that the obstacles it presented could not be overcome by traditional diplomacy alone. He intuited that simple inertia and entrenched self-interest could quickly overwhelm any effort to negotiate agreements on a standard, linear basis.

He instead endeavoured to conduct an intergovernmental heads of state summit on an entirely new model. Instead of focusing solely on governments – or even primarily on governments with a few speakers from universities, the arts and established, non-controversial non-governmental organizations on the side – he decided to conduct a three-year campaign that would demand the active participation of four simultaneously active universes of actors. In what amounted to a diplomatic five-ring circus, he travelled the world appearing at a non-stop sequence of hearings, prepcoms, panel discussions, private audiences, press conferences and interviews.

Strong put together an exceptionally competent staff led by the eminent Indian economist Nitar Desai and an impressive group of earnest advisors. He persuaded a galaxy of others to join him, including media magnate Ted Turner, his enterprising colleague Barbara Pyle, actress Shirley MacLaine and a group of world leaders, former presidents and prime ministers. As his principled colleague and advisor in Brazil, one of its foremost leaders and environmental pioneer Dr Israel Klabin made an indispensable contribution. He also enlisted the support of the scientific community through the cooperation of the International Council of Scientific Unions (ICSU) to ensure that the conference agenda was guided by the best of science, and he set up a separate foundation to finance the cost of all their arrangements.

At each stop, in each capital, he would meet not only with representatives of governments' foreign ministries, but with ministries of the environment, of energy, forests, fishing, oceans, rivers, water, natural resources, and finance and development, too, as well as civil society representatives.

He told 'Southern' developing governments that this was their chance to gain leverage in international economic negotiations because they finally had something that 'Northern' developed governments wanted – their vast and still unpolluted ecological resources whose conservation could help save the environmental health of the planet. He told developed-country governments they had an opportunity to define the parameters of environmental issues for 50 years, to effectively address problems their voters were concerned about, and to help assure a steady access to suppliers and growing export consumer markets for their businesses.

He met with business owners and CEOs, and told them as a former CEO himself that by getting on the bandwagon early, they could reap the benefits of an environmentally-friendly image with consumers, an increased profile with investors, and a seat at the table with regulators. He met with representatives of trade unions and suggested that this was an opportunity to advocate for both the environmental safety of and improved economic standards for workers. He met with environmental NGOs – and leaders of development, social equity, women, youth, religious, agricultural and indigenous peoples. He urged them to get involved, knowing that the conference needed the public visibility and pressure they could bring from the *outside*, and promising that they would have real access *inside* the process to advocate for their issues. He gave special attention to indigenous peoples, and arranged a separate summit of their leaders in conjunction with the conference.

He met with the press – often – and advised them that there was as major story brewing here. 'Rich countries, poor countries, environmentalists, labour leaders, businesses, farmers, even indigenous and church groups – all were going to Rio for the biggest, most colourful, most visionary summit of the century.' He explained the substantive reasons why the media should give a high priority to covering the conference.

Stakeholder engagement up to Rio

It is difficult to imagine there could be a time when stakeholders were not a formal part of the process in the global sustainable development and environment arena. But there was. The emergence of serious stakeholder involvement in global processes was started by the Stockholm Conference, and encouraged at Brundtland Commission hearings where the Commission had reached out to non-state actors for input to the drafts of the Commission report.

The setting up of the Centre for Our Common Future to promote interest in the inputs to and outcomes from the Brundtland Commission, and later UNCED, provided critical institutional assistance. Under the capable leadership of Chip Lindner, the Centre started immediately to reach out to stakeholders through their newsletter *Network 92*. By Rio it was being posted out to over 125,000 people in five languages. This helped to get the message out and to engage a wide range of stakeholders from around the world. Chip Lindner explained the work of the Centre:

> The Centre for Our Common Future also established a network of working partners around the world. Originally we targeted 100 key global networking groups such as the International Chamber of Commerce, the Global Tomorrow Coalition etc. We got them to associate with the Centre for Our Common Future publicly as working partners by way of making a public commitment to further the concept of sustainable development.
>
> (Lindner interviewed in Lerner, 1991: 240)

One of the most important pre-meetings for NGOs took place in March 1990 in Vancouver.

> This had been intended as a review by COCF of its preparations for UNCED. However, after a meeting between Lindner and Maurice Strong, newly appointed as Secretary-General for UNCED, it was agreed that the structure of the meeting should be altered. Lindner later recalled: 'I went to him and said we would be happy to provide our assistance and support to mobilize in the broader constituencies, but we could not work solely from an NGO point of view'. The Centre brought together 152 of its working partners from 60 countries, representing the broader constituencies they had endeavoured to involve in dialogue on the Brundtland Report. These included industry, trade unions, women, youth, media, and NGOs. Half of those present were from developing countries, half from developed.
>
> (Lemer, 1992)

This Vancouver meeting endorsed the COCF call for broad participation by all sectors of society in the UNCED process and gave the Centre a mandate to extend its work in this area. It is worth noting the close correlation between Lindner's account of the conclusions reached and the positions subsequently taken on these issues by Maurice Strong and the UNCED Secretariat:

> At the time the whole question of who could participate in the UNCED process had not been established. First, out of the meeting came a very strong call for broad participation. There was also a recognition that we had to find new mechanisms for resolving problems. Second, participants in the Vancouver conference called for at least 50 per cent participation from developing countries and women in all strategizing and planning for the UNCED conference at all levels – national, regional, and international. Third, it was recognized that the development side of the environment/development nexus was extremely weak in the proposed agenda for 1992; and that the inter-sectorial or cross-cutting issues, as they are now called, were elementary issues that had to be seriously addressed. And fourth, it was decided that the Centre for Our Common Future should call a meeting of heads of institutions to get a mandate to play some kind of focal point role in 1992.
>
> (Lemer, 1992)

A second meeting was held in Nyon, Switzerland in June 1990. Over 100 representatives from different sectors and geographical regions were represented. An International Facilitating Committee (IFC) was established to 'serve as a focal point for independent sector efforts for Rio. Its role would be to organize the Global Forum in Rio, support access to the negotiations and organize facilities for the stakeholders at the preparatory meetings.

The major International NGO Roots of the Future conference was held in Paris in December 1991 – through the support of the government of France and organized by the Environmental Liaison Centre International (ELCI) – in preparation for the Earth Summit saw the development of global stakeholder networks in areas that had none before committed to sustainable development and the strengthening of others.

Over 800 non-governmental organizations and citizens' organizations attended the conference with nearly three-quarters coming from developing countries. The main outcome document was Agenda ya Wananchi (ELCI, 1991) which in Swahili means 'Sons and Daughters of the Earth'. The document included calls from NGOs for governments to support participatory democracy, a powerful and effective UN for governments to reduce their military spending by at least a half a request to developed countries to reform the world's trading system; and to increase financial flows to the developing countries and the newly independent countries of the former Soviet Union.

The meeting also called for building global NGO coalitions to work together in the 'struggle for global justice and sustainability'. They also called for a commitment to 'campaign against all those national and international organizations and interests who disregard the imperatives of justice and sustainability', 'the development of equitable and sustainable natural resource management systems and technologies', 'a struggle for the empowerment of the socially and ecologically marginalized people'; and a commitment to 'a struggle for women's empowerment and equal status in society' (ELCI,1991).

The Swiss businessman Stephan Schmidheiny, who served as Maurice Strong's principal business advisor, had the task of mobilizing the participation and support of the business community. Schmidheiny set about involving 58 other business leaders through a new organization called the Business Council for Sustainable Development (BCSD). The World Economic Forum put the Earth Summit and the issues it addressed in its highly influential agenda.

The International Chamber of Commerce (ICC) produced a Business Charter for Sustainable Development. ICC endorsed the Charter at their pre-Rio Conference. The charter put no obligation for any company that signed it to do what it said and there was no independent audit proposed for the charter. There had been a proposed chapter 41 of Agenda 21 'Sustainable Development and Transnational Corporations' put forward by the UN Centre on Transnational Corporations, this had not even survived to the negotiating table, so the Charter was cited in Agenda 21 as a valuable code of conduct, 'promoting best environmental practice'. Within two years the Centre had been closed and its activities submerged into UNCTAD.

To champion the rights of women in the Earth Summit process, former US Congresswoman Bella Abzug and feminist activist and journalist Mim Kelber created the Women's Environment and Development Organization (WEDO) in 1991. WEDO organized the World Women's Congress for a Healthy Planet, bringing together more than 1,500 women from 83 countries to work jointly on a strategy for Earth Summit.

The science community gathered in November 1991 for the Vienna International Conference on an Agenda of Science for Environment and Development into the Twenty-first Century (ASCEND 21) organized by the International Council of Scientific Unions (ICSU) together with the Third World Academy of Sciences.

The Trade Unions met in March 1992 in Caracas, Venezuela, for the World Congress of the International Confederation of Free Trade Unions to input to Rio.

To help local government input to the Summit process the International Council for Local Environmental Initiatives was founded in 1990. The Council was established when more than 200 local governments from 43 countries convened at our inaugural conference, the World Congress of Local Governments for a Sustainable Future, at the United Nations in New York.

It met just prior to Rio in Curitiba, Brazil, hosting a World Urban Forum in preparation for UNCED, organized jointly by UNDP and Curitiba City Council.

Parallel to stakeholder preparation, the formal process had started with a first organizational preparatory meeting in Nairobi in August 1990. For the first time Canada, Norway and the UK and some other countries included NGOs on their delegation.

What followed were regional meetings that offered the first opportunities for countries to discuss priorities for the Rio Summit.

Perhaps the most advanced as far as stakeholder engagement was concerned was the Bergen Conference. The conference was at ministerial level; it also for the first time offered a structured attempt to promote interaction between NGOs and government delegations. The conference coined the term 'independent sector' which continued in use until 'Major Groups' came in as Agenda 21 chapters.

Following up on this, Strong appointed Pierre-Marc Johnson, an influential former Premier of Quebec, Canada, as senior advisor. He played an important part in engaging the involvement of developing countries in the Summit.

The Brundtland Bulletin, produced by the Centre for Our Common Future stated that:

> For most observers, what really mattered about Bergen was that it saw the emergence at an international level of a new and unique participatory process in which bodies representing the 'independent sector' (i.e. industry, trade unions, the scientific community, youth, and non-governmental organizations concerned with environmental issues) not only conducted their own parallel conferences, but participated with the ministerial delegations in the quest for the broadest possible consensus. The 'Bergen Process' of consensus-seeking between independent and official channels had been evolving over the two years in which Bergen was in preparation, and seems set to become the model for 'the 1992 process', as we now move towards the all-important UN Conference on Environment and Development in Brazil.
>
> (Brundtland Bulletin)

At Bergen there were two important developments which would frame the landscape of stakeholder involvement in intergovernmental meetings from here on. Stakeholders started to work in groups such as NGOs, women, trade unions and youth. And governments started to realize that for their decisions to be accepted by society they needed the involvement of stakeholders. Stephen Collett of the Quakers Office to the United Nations outlined the shift in stance which led to the acceptance of participation by such organizations:

> That NGOs have an important contribution to make is now generally accepted; the question under debate is how to channel contributions into the intergovernmental process, and particularly how the invitations to and statements of non-consultative NGOs will be handled. While a majority of statements in this debate were rosy on the role of NGOs – the USA, for example, read a list of important US national organizations which would not have consultative status but have useful expertise – the greatest shift is in the G-77. From having been somewhat discomfited by the whole idea, the mainline G-77 statements now called for formulas to bring 'broad-based and balanced' NGO participation to the Conference and its preparatory process,

laying weight on the need for equal representation of groups from their regions with those of the North. This marks something like a 170 degree shift from their strict stand of earlier discussions, aimed at limiting participation to those they couldn't keep out – those in consultative status.

(Collett, 1991)

When Maurice Strong met with representatives from the 'independent sector' after the Bergen Conference he 'stressed his support for the principle of broad representation and participation'.

The UNCED Secretariat prepared guidelines for NGO participation that recommend that NGOs, as well as groups from a broad spectrum of society, be brought into the official process.

The outreach to stakeholders was also to be undertaken at the national level where he suggested that governments, when drafting their national reports, should involve all sectors of their society.

Chip Lindner of the Centre for Our Common Future commented:

we have to find a way to move from confrontation through dialogue to cooperation; and we have to get all the players at the table. It is no longer good enough to be critical. Each of us has to accept a share of the responsibility to do something. And we all have to have the humility to recognize that our solutions are not necessarily the only ones or ultimately the right ones. The world works inter-relatedly and we have to work inter-relatedly.

(Lindner, 1992)

Chapter 3

An Earth Summit

Parents used to be able to comfort their children by saying, 'Everything's going to be all right.' 'We're doing the best we can.' And, 'It's not the end of the world.'

But you can't say that to us any more. Our planet is becoming worse and worse for all future children. Yet we only hear adults talking about local interests and national priorities. Are we even on your list of priorities? You grown-ups say you love us, but we challenge you to make your actions reflect your words.

(Severn Cullis-Suzuki, 12-year-old Canadian
NGO Plenary Address to the Earth Summit, June 1992)

The achievement of the Earth Summit in Rio de Janeiro in 1992 was extraordinary. Producing Agenda 21 – the blueprint for sustainable development for the first part of the twenty-first century – was by itself an unprecedented accomplishment. The Rio Declaration provided a set of principles to guide the journey. And the launching of two extremely significant conventions – on biodiversity and climate, as well as the Forest Principles – put those theories into action.

It is important to remember that this was accomplished while the United States played an unhelpful role in key negotiations. The US successfully ensured that there were no targets agreed in the climate change agreement and refused to sign the biodiversity convention at all.

The Conference preparations had begun amidst considerable optimism that the end of the Cold War might yield a 'peace dividend' that could fund the necessary movement towards sustainable development. But the meeting itself occurred in the aftermath of the first Gulf War, which had helped to spark increased energy costs and a new economic recession, that undermined the capacity – and the willingness – of governments to agree to anything close to such funding. In spite of these difficulties Rio was, in fact, one of the most comprehensive intergovernmental negotiations on any issue, to date.

The Earth Summit opened on 3 June, and the next ten days would set the international environmental and sustainable development agenda for the next twenty years and the hopes for that generation.

Outcomes

The Summit was held in a venue which became known as 'Riocentro', a seemingly vast aircraft hangar, 40 kilometres outside the city. It was attended by the governments of 178 nations, 108 of which sent their heads of state or government. At the time, that represented the largest number of heads of state or government ever to attend a United Nations conference or summit. In addition, 2,400 representatives of non-governmental organizations (NGOs) and roughly 10,000 journalists were accredited to the official Summit itself.

Estimates of the stakeholders attending the parallel 'Global Forum' events facilitated by the Centre for Our Common Future in Flamenco Park varied from 35,000 to 50,000.

The Earth Summit and its multitude of side events constituted the largest global gathering on sustainable development ever, and set the model for a decade of international summits on broad 'new' issues, ranging from women to children to cities to food security. It also resulted in the following specific agreements:

- The Rio Declaration on Environment and Development: A set of 27 principles intended to guide future sustainable development throughout the world.
- Agenda 21: A 40-chapter blueprint for action in the twenty-first century that would continue social and economic development, increase conservation and management of resources, strengthen the role of major groups, and provide the means for those activities' implementation.
- The United Nations Convention on Biological Diversity (CBD): The Convention had three main goals:
  1 conservation of biological diversity;
  2 sustainable use of its components; and
  3 fair and equitable sharing of benefits arising from genetic resources.
- The United Nations Framework Convention on Climate Change (UNFCCC): An international environmental treaty to provide a framework for stabilizing greenhouse gas concentrations in the atmosphere at a level that would prevent dangerous anthropogenic interference with the climate system.
- The Forest Principles: The informal name given to the 'Non-Legally Binding Authoritative Statement of Principles for a Global Consensus on the Management, Conservation and Sustainable Development of All Types of Forests'. The Forest Principles are to guide countries towards sustainable management of their forests.

Other outcomes included:

- The Global Environmental Facility: Established in 1990, as the fund for the Montreal Protocol, it was given added responsibilities for biodiversity, climate change and international waters after Rio.

- The Convention to Combat Desertification (CCD): A convention heavily lobbied for by African countries, it was negotiated within two years of the Earth Summit.
- The United Nations Commission on Sustainable Development (UN CSD): Established after UNCED as a functioning Commission of the United Nations Economic and Social Council. The CSD's responsibilities were to monitor the implementation of the Rio agreements and to negotiate future policy commitments.

Finally, the Rio process set new models for expanding intergovernmental decision making, by providing space for: the active participation of a vast range of civil society organizations – businesses, trade unions, NGOs, woman's organizations, scientists, and local governments –inside the official negotiating process:

- The Global Forum in Rio, a gathering of stakeholders sharing ideas, campaigns and good practice.
- The creation of new stakeholder institutions established in the run up to Rio and one significant one after the Summit.
- The building of public awareness and political momentum – the Earth Summit's massive success at reaching the world's public via global media vastly raised support for environmental and development issues among the international public.

The Rio Declaration

The Rio Declaration, although a politically important document, wasn't what the secretariat had hoped for. The original expectation had been for an 'Earth Charter' which would reaffirm and build on the Declaration of the Stockholm Conference, the IUCN Covenant and other similar values statements agreed upon in previous intergovernmental processes. The Earth Charter would have set out the basic principles to guide the conduct of nations and peoples towards each other and towards the planet – a kind of Universal Declaration of Human Rights for the Earth.

Chairing these negotiations throughout had been Bedrich Moldan the Czechoslovakian Environmental Minister. As he recalls:

> One of the most memorable was in Washington DC where I was invited to the lunch at State Department that was attended by seven(!) lawyers. At the end of the meeting I was very pleased by their assurance that while they generally thought that any document of such kind as Earth Charter is superfluous, in essence they will not block its acceptance. (In fact, the US later stuck faithfully to this position.)
> (Moldan, 2011)

The blocking came from Pakistan as head of G77 who could not accept the Earth Charter text. Tommy Koh handpicked a group of about 16 people (informal 'Friends of the Chairman') to then develop an alternative, the Rio Declaration.

The text of the Declaration was produced in a short time. It incorporated some of the original Earth Charter elements but mostly built upon previous 'agreed texts', above all the Stockholm declaration. Some of the elements were quite new including the famous Principle 7 on 'common but differential responsibility' of states.

In the absence of action by governments, the Earth Council – a new, high-profile NGO formed after the Summit – would continue to promote the need for an Earth Charter, and would develop it further in preparation for the 2000 Millennium Development Summit.

Nonetheless, the Rio Declaration has had its own positive impacts. What has evolved since 1992 is that a number of its significant principles have found their way into the domestic law of many countries. As with the Stockholm Declaration, what was once a set of 'soft law' agreements have now been integrated into 'real law' and are having an active effect on national policies and actions.

A most recent example involves the Declaration's Principle 16, colloquially known as 'polluter pays':

> National authorities should endeavor to promote the internalization of environmental costs and the use of economic instruments, taking into account the approach that the polluter should, in principle, bear the cost of pollution, with due regard to the public interest and without distorting international trade and investment.
>
> The Rio Declaration (United Nations, 1992c)

In the 2010 Gulf of Mexico oil spill, which mostly affected the United States, multinational petroleum giant BP agreed to pay billions of dollars in compensation to companies and individuals following intense negotiations with the US government, based on an agreed determination of corporate responsibility. The principle has also become an underlying basis of climate change negotiations, in which developed countries acknowledge that damages caused by their industrial activities require compensation to those developing countries incurring the negative impacts.

Another example of a Rio Declaration principle that has taken on legal form is Principle 10, which promotes:

> the access to information concerning the environment that is held by public authorities, including information on hazardous materials and activities in their communities, and the opportunity to participate in decision-making processes. States shall facilitate and encourage public awareness and participation by making information widely available. Effective access to judicial and administrative proceedings, including redress and remedy, shall be provided.

Principle 10 has become the basis for a Europe-wide regional convention known as the Aarhus Convention, adopted in that Danish city on 25 June 1998 at a ministerial-level conference of the 'Environment for Europe' process. The Aarhus Convention:

- links environmental and human rights;
- acknowledges that present generations owe an obligation to future generations;
- establishes that sustainable development can be achieved only through the involvement of all stakeholders;
- links government accountability and environmental protection;
- encourages interactions between the public and political authorities in a democratic context.

(UNECE, 1998)

Although it is a regional convention, it is open to global ratification. To date, no developing countries have signed the convention. The World Resources Institute has taken up the challenge to work at making Principle 10 operational in developing countries.

Perhaps one of the Rio Declaration's weakest areas was in the area of military conflict. It did not address reducing weapons of mass destruction, unlike the Stockholm Declaration of 1972, whose Principle 26 had said:

> Man and his environment must be spared the effects of nuclear weapons and all other means of mass destruction. States must strive to reach prompt agreement, in the relevant international organs, on the elimination and complete destruction of such weapons.

Perhaps governments – considering the breakup of the Soviet bloc only two years earlier – felt they didn't need to address military matters at that time. The closest expression of the need for addressing conflict can be found in the Principle 25 formulation that 'Peace, development and environmental protection are interdependent and indivisible.'

Agenda 21

> Humanity stands at a defining moment in history. The world is confronted with worsening poverty, hunger, ill health, illiteracy, and the continuing deterioration of ecosystems on which we depend for our well-being. The disparities between rich and poor continue.
>
> The only way to assure ourselves of a safer, more prosperous future is to deal with environment and development issues together in a balanced manner. We must fulfill basic human needs, improve living standards for all and better protect and manage ecosystems. No nation can secure its future on alone; but together we can: in a global partnership for sustainable development.

(Agenda 21, 1992: Preamble para 2)

Agenda 21 reflects a global consensus and political commitment at the highest level of government on development and environmental cooperation. The agenda deals with both the pressing problems of today and the need to prepare for the challenges of the next century.

> It recognizes that sustainable development will have to be delivered not only by government but will need the involvement of all relevant stakeholders. Governments will have the responsibility for leading through the development of national strategies, plans and policies. The efforts of nations need to be linked by international cooperation through such organizations as the United Nations. The broadest public participation and the active involvement of the non-governmental organizations and other groups should also be encouraged.
>
> (Agenda 21, 1992: Preamble)

Agenda 21 is one of the most comprehensive documents created by any intergovernmental process. Its 40 chapters provide a blueprint for the way humans need to live on their common planet to be able to assure the viability of its ecosystems and their own survival in the beginning of the twenty-first century. In a breathtakingly ambitious effort, it organically integrates such diverse factors as the necessity for economic vitality, providing resources for the poor, the responsibilities of national governments, international governmental organizations, funding, activists, educational, religious organizations and local authorities. It carefully assesses the roles of major stakeholders in society, industry, trade unions, local government, women, the needs of youth and the rights of indigenous peoples. And of course, it itemizes the environmental state of forests, oceans, freshwater, the atmosphere, ecological systems and the global climate, describes the multidimensional interactions between them and economic and social dynamics, and prescribes remedial strategies to address all.

Although the draft was weakened throughout the negotiating process, the agreed text still stands as the most far-reaching and, if implemented, the most potentially effective programme of international action ever sanctioned by the international community. It was not seen as the final and complete programme of action, but rather one that should continue to evolve as new information and new insights arrived. What it failed to do was to deal with the interlinkages between issues effectively.

Agenda 21 was organized into four areas:

- Section 1: Social and Economic Dimensions: deals with combating poverty, changing consumption patterns, promoting health, demographic dynamics and sustainability (population change), and sustainable human settlement development.
- Section 2: Conservation and Management of Resources for Development: includes atmospheric protection, combating deforestation, protecting fragile environments, conservation of biological diversity (biodiversity), freshwater, oceans chemicals and control of pollution.
- Section3: Strengthening the Role of Major Groups: deals with the nine sectors of society that were identified to have a role in implementing Agenda 21.

- Section 4: Means of Implementation: includes capacity building, education for sustainable development, finance and financial mechanisms, science, technology transfer and international institution.

Each chapter and chapter sub-section was built around a common structure for addressing its issue. It consisted of

- basis for action
- objectives
- activities
- means of implementation.

This created a consistent narrative of why an issue was being addressed; what the goals would be in addressing it; what set of activities would be initiated by not only the UN, but by governments and stakeholders; and what was needed to achieve the objectives (e.g. finance, capacity building, education, or technology transfer). Such a structure was unique in a UN document.

There were key missing chapters of Agenda 21, perhaps the most obvious being a chapter on energy. Effective lobbying by oil-producing nations had ensured that there was no chapter in Agenda 21 addressing the impacts of fossil fuel consumption and the need to shift to environmentally friendly energy strategies and technologies. It was the only issue on which the secretariat had informally suggested that there might be a majority vote – therefore breaking the rule of decision by consensus. In the end that vote did not happen at Rio, and the battle over oil and energy was left to the ensuing negotiations on the Convention on Climate Change. The gap in Agenda 21 on energy issues would be rectified in 1997 at the Rio+5 meeting of the General Assembly.

Agenda 21 is one of the first UN documents to attempt to address the impacts on the planet of the consumption patterns that had developed in the latter part of the twentieth century. Not every country was willing to address its levels of consumption. In an (in)famous statement by the first President George Bush, the US leader during the 1992 Earth Summit declared that:

> The American Way of Life is non-negotiable.
>
> (Bush, 1992)

Emotionally powerful slogans can be highly effective – no matter their factual inaccuracy. The statement was subsequently reiterated verbatim by US Vice President Richard Cheney, in 2004, in rejecting any cap on carbon emissions implemented to reduce global warming.

Ultimately the Agenda 21 policies are meant to address changing consumption patterns, population and resource use so that people in all countries might live within what are now called planetary boundaries with fairness and an increasing degree of equity.

Like the Stockholm Action Plan, Agenda 21 puts eradication of poverty at the centre of discourse. The 'Rio deal' established that developed countries had a much larger responsibility for causing damage to the global environment than had poor countries, and therefore a much larger responsibility for cleaning up that damage. This is the implicit understanding expressed through Principle 7 of the Rio Declaration, which recognizes that 'States have common but differentiated responsibilities'.

To enable developing countries to move towards a more sustainable form of development, there needs to be a transfer of technology, a building of capacity and an increase in funding. In response to a query, the conference secretariat estimated that it would cost US$600 billion per year to implement Agenda 21, with US$125 billion of that moving from developed to developing countries. At the time of the summit, total global Official Development Aid (ODA) was only US$60 billion – making the secretariat's estimate seem to many governments like an overwhelming and unaffordable amount. In retrospect, perhaps the unrequested information that might have been provided was what would be the costs to *not* implement Agenda 21.

The implementation of Agenda 21 was seen as requiring 'new and additional' financial resources. The Conference Secretary-General, Maurice Strong, had suggested the establishment of an 'Earth Increment' – an added provision of dedicated funds beyond existing ODA. This was not a suggestion appreciated by the US delegation, which was averse to any action that would require additional governmental funding. Strong suggested a series of strategies that could address American concerns. These included new taxes, user charges, emission permits, and citizen funding, all of which were based on the 'polluter-pays' principle. He also reminded governments that ending the waste of funds caused by existing subsidies to non-environmentally-sound activities could alone provide all the funding necessary and urged them to see it as an indispensable investment in environmental security.

At the end, though, there wasn't a grand deal on finance at Rio, but a finessed diplomatic agreement.

The chair of the G77 at the time of the Summit was the government of Pakistan, represented by Ambassador Jamsheed Marker, who therefore spoke as chairman of the G77 bloc. He adroitly obscured the reality that no new money was on the table by indicating that developing countries would 'acknowledge in principle' that new and additional financial resources would be forthcoming after the Summit once developed countries' economies were in better state. That would leave these negotiations for a future time. The head of the US delegation, Buff Bohlan, seemed to strongly affirm the agreement:

> There is no question that developing countries and countries in transition must have new resources. I would like to make it absolutely clear that the United States is committed to working with other industrial countries to mobilize new and additional resources for a new partnership.
>
> (Bohlan quoted in Strong, 2001: 213)

This was an extraordinarily strong statement. Whether it indicated a genuine support for providing more future government funding, or was merely a subtle signal of his government's future policy of 'mobilizing' resources by enabling extensive corporate investment is open to debate.

French President François Mitterrand did pledge that France would meet the 0.7 per cent target by the year 2000. But, in fact, its ODA contribution actually fell from 0.63 per cent in 1992 to 0.31 per cent by 2000, another broken promise.

The issue was then to be revisited in the new Commission on Sustainable Development that would be formed following the report of the Conference, by the General Assembly, and would convene in 1993. The terms of reference for the Commission included an annual review of ODA, new financial commitments, technology transfer and capacity building to implement Agenda 21.

Ultimately, Agenda 21's finance strategy was an attempt to secure unprecedented funding support from the UN system, from governments, from financial institutions and from all stakeholders. As such, it was a bold and ambitious attempt to build a vehicle capable of taking all nations on a cooperative journey along a path to a sustainable world.

The key word was 'cooperative'. As one of the leading organizers said in preparing for the Conference:

> we have to find a way to move from confrontation through dialogue to cooperation; and we have to get all the players at the table. It is no longer good enough to be critical. Each of us has to accept a share of the responsibility to do something. And we all have to have the humility to recognize that our solutions are not necessarily the only ones or ultimately the right ones. The world works inter-relatedly and we have to work inter-relatedly.
>
> (Lindner, 1992)

In real terms the 'money on the table' was pitiful. The United States early announced an extra US$150 million in the area of forests. Meanwhile the European Commission pledged US$4 billion and Japan promised to increase its ODA from US$3.1 billion to US$7 billion over the next five years. By the conclusion of the Summit, however, only US$6–8 billion had been committed – far from the estimated requirement of US$125 billion per year. And there was no certainty that even this was new money, and not funds recycled from previous commitments.

What it didn't address was the real structural economic issues such as trade and debt which were impacting on sustainable development. The NGOs were the only ones who were openly arguing that the free market and what became known as globalization should be subordinate to sustainable development.

The attempt by the NGOs and some developing countries to put forward some form of regulation on multinational corporations did not find much support in a time period which was taking away regulation not adding it.

The Convention on Biological Diversity

As the Earth Summit convened in 1992, following two and a half decades of well-publicized loss of habitat and species in their own countries, governments in developed countries should have been motivated to take international action to preserve natural habitat by their own failures. But the immediate impetus for their concern was the more recent alarm about the destruction of tropical rainforests.

It was, in fact, the government of the United States, under then President Ronald Reagan, which first suggested a Convention on Biological Diversity at the UNEP Governing Council in 1987. The US suggestion was to:

> rationalize arrangements under existing international conservation agreements and their localized secretariats, with a view to bringing everything together under an 'umbrella' convention.
>
> (McConnell, 1994: 5)

Initially the US did not want to see biotechnology included in the negotiations on the convention, but it eventually agreed (at the 1989 UNEP Governing Council) to work on 'an international legal instrument' including social and economic issues and 'the use of genetic resources in biotechnology development'. This laid the foundation for negotiations that would start in 1990. They were not originally planned to be completed in time for Rio.

That it was being negotiated at all reflected a growing realization that the ecosystems that humans lived in were not just resources to be exploited, but the fundamental life-support systems of the planet. The economic systems that had always viewed forests as infinite resources that could be cleared for agriculture or timber now needed to start to consider those ecosystems as having ecological and financial value, and therefore no longer as infinite and 'free'.

The goal of advocates of a biodiversity convention was to formalize agreement that what had typically been considered a free resource for plunder by pharmaceutical and extractive industries, could no longer be seen that way.

Developing countries and the representatives of indigenous peoples asserted that they had sovereign control, and demanded therefore that there should at the least be profit sharing from any commercial use. It was perverse, they argued, that their people should not only be denied a benefit from products derived from the plants and often the practices of their regions, but should then also be denied the right to produce those products for themselves.

The response from some developed countries, in particular the United States, was that their companies deserved continued free access to resources, while demanding that the intellectual property of corporate developers should be protected. In his statement to the plenary, President Bush presented a fierce rebuttal, and counter-attacked the financing provisions of the convention:

> I didn't come here to apologize ... The financing scheme will not work ... [The convention] threatens to retard biotechnology and undermine the protection of ideas.
>
> (McConnell, 1994: 111)

As political pressure built on the Bush Administration, from NGOs and other governments to agree to at least one major treaty, the US negotiated and agreed to accept modifications in the draft language. However, those changes in the convention – watering down provisions regarding ownership of genetic resources – led the Third World Network, the respected Southern-based NGO, to advise developing countries not to sign.

> Nothing in the convention indicates respect for the rights of indigenous people.
> (Corpuz quoted Weismann, 1992)

Ironically, despite having won concessions at the convention, the US still did not agree to sign. Even friends of the US had tried to persuade them to sign. Ros Kelly, leader of Australia's UNCED delegation:

> It is disappointing that the United States has so far not indicated they will sign. I urge them to do so, if not in the interest of the living things of the planet, then in their own self-interest. ...We are the dominant species on the planet, but we depend for our continued survival on all the others.
> (Kelly quoted in Weismann, 1992)

The UK looked like it also wasn't going to sign. But a last minute publicity campaign, run by the Worldwide Fund for Nature (WWF), succeeded in getting the British government to shift its position. In fact, the Convention on Biological Diversity was signed by more countries than was the convention on climate change, a total of 156 plus the European Union.

The Convention has three main goals:

1 conservation of biological diversity (or biodiversity);
2 sustainable use of its components; and
3 fair and equitable sharing of benefits arising from genetic resources.

The United Nations Framework Convention on Climate Change

It's been a long journey to get governments to act seriously on climate change. Perhaps the most hopeful meeting prior to the Earth Summit had been in Toronto, in October 1988 at the 'World Conference on the Changing Atmosphere: Implications for Global Security'.

It was there that many developed countries' governments had pledged to stabilize their carbon dioxide emissions by the year 2000, and then to achieve a 20 per cent reduction from 1990 levels by 2005. The Toronto meeting was also critical in the establishment of the Intergovernmental Panel on Climate Change (IPCC). While the first report of the IPCC – issued in time for the Second World Climate Change Conference in 1991 – had stated that there was not 'conclusive' scientific evidence that emissions of carbon dioxide and other greenhouse gases from human sources were changing the filtering mechanism of the Earth's atmosphere, it concluded that emissions were 'likely' to produce warming of the planet's climate, accompanied by increasing climate turbulence.

In the United States, the National Academy of Science Committee on Science, Engineering and Public Policy published a report titled Policy Implications of Greenhouse Warming, which seemed to say that the US could adapt to any changes in climate, but only if it continued on a path of strong industrial growth:

> A final limit, and a common one, is money. Adaptations like furnaces and air conditioners, sea walls and canals, take money. Resources for such investment require continuing ability to generate wealth.
>
> (National Academy of Science, 1992: 503)

Major industrialized economies like Australia, Canada and Japan had sided with the US. Even the post-collapse Soviet Union – though now taking a strongly 'green' position on a broad range of issues – could not accept limits on emissions if it meant reducing economic production. The oil-producing nations (OPEC) – despite their membership in the G77 bloc of developing countries – were adamant that there be no limitations on production of petrochemical energy.

Environmental NGOs supported the position of most developing nations, arguing that slowing climate change by strictly reducing greenhouse gas emissions was both an ecological and moral necessity.

One of the NGO Treaties negotiated at the Global Forum parallel NGO event, the Alternative Non-Governmental Agreement on Climate Change, called for:

> industrialized countries to reduce carbon dioxide emissions by at least 25 percent from 1990 levels by 2005.
>
> (International NGO Forum, 1992b)

The European Commission Environment Commissioner was so angry about the role that the US was playing he refused to attend the Earth Summit. He said that the Earth Summit:

> was intended to take decisions, obtain precise and concrete commitments to counteract tendencies that are endangering life on the planet.
>
> (Weissman, 1992)

The European Commission announced that the 12 countries had an internal agreement to reduce greenhouse gas emissions to 1990 levels by 2000.

Maurice Strong argued to governments that preventative action was feasible and could produce many benefits. In particular there would be substantial improvements in energy efficiency and conservation with existing technologies.

The developed and developing nations split, as they had in Stockholm in 1972. At that time, it was over general principles of responsibility for environmental degradation. This time, it applied to the specific issue of climate change. Developing countries arrived at the 1992 negotiations in Rio with a clear belief that because global warming was overwhelmingly caused by developed countries, it was those wealthier nations' responsibility to reduce their own emissions first. They then needed to make financial and technical resources available to developing countries so that they could follow a more sustainable path, and not repeat the energy mistakes of industrialized counties.

Given the impasse, it was clear from early on that the best achievement possible would be to agree on pursuing a framework convention, which would leave the negotiations on specific targets and timetables to be accomplished later, through a protocol to the convention – similar to the way negotiations on destruction of the ozone layer were being addressed.

Ultimately, approving the framework convention was only a first step for dealing with climate change – and not a sufficient one by itself. Stabilizing the gaseous composition of the atmosphere was arguably the most urgent environmental problem the world would face in the 1990s. Many scientists were reporting that carbon emissions must be cut by at least 60 per cent just to put the global warming trend on hold.

Yet the agreement signed at Rio de Janeiro established neither targets nor timetables. Governments were not yet prepared to take the actions needed. As a result, critical time was lost. It took until December 1997 for the Kyoto Protocol to be agreed upon, and until 2005 to ratify the target of a total 5.2 per cent emissions reduction from 1990 levels by the year 2012.

The Forest Principles

Forests were dealt with in two places in a chapter in Agenda 21 and in what became the Forest Principles.

Perhaps one of the most disappointing outcomes from the 1992 Summit was the negotiations on forests. Originally the expectation was for this to be the subject of a third major convention. The push for a convention was led in 1990 by a group of developed countries including Canada, France, Germany, Italy, the UK and the US. As with negotiations on biological diversity, the original motivation for a convention was to deal with the world's tropical rainforests.

The US tried to take a lead in promoting forests to balance its negotiating in the other areas. The US$150 million increase in its international assistance for forests announced by the US was seen by many NGOs as an insult. Bill Mankin of the Sierra Club, one of the largest US NGOs, said:

> This initiative is a year late, at least three and a half billion dollars short, and is practically guaranteed to offend developing countries.
>
> (Mankin quoted in Weissman, 1992)

The principles set out were useful but largely were already accepted and in some cases just reiterated existing guidelines from organizations such as the International Tropical Timber Organization (ITTO), the Food and Agriculture Organization (FAO) and the World Bank.

The attempt to negotiate a convention on forests has continued to elude advocates. Many developing countries fear that the Forest Principles, though themselves representing soft law, could be used to impose restrictions on aid or conditions in future trade negotiations. A convention on forests, which would include financial payment for the role forests play in serving as a 'sink' for carbon dioxide storage, took another 18 years to come to an agreement not as a separate convention but as part of the climate convention.

Most recently, a different approach to preserving forests has been attempted – through the Climate Change Convention. Reducing Emissions from Deforestation and Forest Degradation (REDD+)', agreed upon at the climate negotiations' COP 16 in 2010 in Cancun, paves the way for forest preservation funding, though it did not escape criticism from indigenous peoples and NGOs:

> Mayan elders expressed that it is unethical and not in accordance to their traditions and ancestral ways to participate in the REDD program that would pay them money in an offset program that allows polluters to continue to pollute, resulting in a program that would cause the warming of the Mother Earth and not for their stewardship of their forests.

> (Goldtooth, 2010)

Forests were to continue to be debated in the coming years through institutions such as the Intergovernmental Panel on Forests, the International Forest Forum and ultimately the permanent body the UN Forest Forum.

Global Environment Facility

The Global Environment Facility (GEF) started as a pilot funding mechanism based on proposals made by France and Germany in 1989. It was established just before Rio in November 1991 as a $1 billion programme within the World Bank, in partnership with UNDP and UNEP. GEF initially aimed:

> to assist in the protection of the global environment and to promote environmental sustainable development. The GEF would provide new and additional grants and concessional funding to cover the 'incremental' or additional costs associated with transforming a project with national benefits into one with global environmental benefits.

> (GEF, 2011)

Initially, the Fund was set up to support projects in four focal areas: climate change, biodiversity, ozone depletion, and international waters. During the Rio process, developing countries called for a 'green fund' to support national environmental actions. In Rio, it was agreed that the GEF would become the funding mechanism for global environmental actions provided that its governance is democratic, its decisions are transparent, and without imposing new conditionalities. After Rio, both the climate change and biodiversity conventions accepted the GEF as their financial mechanisms.

This was still far from what developing countries had asked for which was a 'green fund' which would have equal votes between developed and developing countries. They objected to the GEF being the funding mechanism as it had a system of voting that was oriented to the donors, which they saw as undemocratic.

Martin Khor director of the Third World Network summed up well when he said:

> In this situation, aid becomes a symbol rather than a solution. And the symbolic importance of UNCED's reliance on the GEF instead of a Green Fund was that it signified the power being appropriated by the industrialized country-controlled Bretton Woods institutions – the World Bank, the International Monetary Fund and the General Agreement on Tariffs and Trade – at the expense of UN-affiliated bodies, which are more open to influence from Third World countries and which are run more democratically.
>
> (Khor quoted Weismman, 1992)

After 18 months of intensive negotiations, facilitated by Mohamed El-Ashry, the GEF was restructured and became an independent financial entity governed by a Council with 32 members – 16 developing countries, 14 developed countries, and 2 economies in transition. Decision making became by consensus but if a vote was needed, a double majority system was instituted – a hybrid of the UN's one country, one vote, and the World Bank's one dollar, one vote. The initial contributions to the GEF Trust Fund then were US$2 billion.

Desertification Convention

At the time of Rio around 900 million people lived in the drylands as they were called and the area covered roughly one-third of the world's surface. Dryland areas in Africa were suffering particularly in the 1970s and 1980s from drought causing crop failure.

For some people there is some confusion in the use of the term desertification. Wikipedia defines desertification as:

> the degradation of land in drylands. Caused by a variety of factors, such as climate change and human activities, desertification is one of the most significant global environmental problems.

The UN held its first Conference on Desertification in 1977 and it marked the first attempt of the international community to examine and address the desertification problem. It agreed a Global Plan of Action to Combat Desertification and a funding mechanism to finance anti-desertification. UNEP was given responsibility for the follow up.

By the Rio preparatory meetings, dryland countries were very frustrated by the follow up and the lack of funds to address identified problems. There was a strong call from the African group for a convention on desertification to be negotiated out of Rio as it was too late to negotiate it for Rio. They felt that by doing so there would be a better chance to address some of their key issues such as food security and poverty.

> The OECD countries promised, as a consequence, to start negotiations for a Desertification Convention as a means of keeping African governments engaged in the Rio process.
>
> (Toulmin, 1997)

So desertification became part of the negotiating agreement for Agenda 21 and the two conventions from the African group. Within four years of Rio the Convention would be negotiated and come into force with the required number of ratifications.

The United Nations Commission on Sustainable Development

The scope of Agenda 21 required that there would be a follow up mechanism able to address sustainable development. At the time of Rio, there was no institution that could do that. The preparatory process had identified a number of options for Rio follow up. The most ambitious idea was the transformation of the now inactive UN Trusteeship Council into an agency to take on a role of 'trusteeship' for the Earth. Another was that there is a sustainable development organ of the General Assembly. Such ideas would be revisited a number of times over the next twenty years.

But there was insufficient support for any of the proposals that would have set up a body with significant profile or political clout. And the proposals were complicated as they would need a 'Charter' amendment and governments were very reluctant to re-open the UN Charter.

Even the more weakened option, put forward by the chair of the negotiations, Ambassador Razali of Malaysia, of creating of a UN Commission on Sustainable Development (CSD) as a functioning commission of the UN Economic and Social Council (ECOSOC), was opposed initially by many key countries – including the UK, China, Austria, India, Sweden, Brazil, Japan, Argentina, Australia, Kenya and Norway.

Norway had initially wanted to strengthen ECOSOC but, because of the intervention of Prime Minister Gro Harlem Brundtland herself, they advocated a much more radical position of making the new CSD a subsidiary body of the UN General Assembly supported by Colombia and a number of other developing countries. This is an idea being returned to for Earth Summit 2012.

During the 1980s, developing countries had become concerned about the possible implementation of some form of compulsory national reporting, and were reluctant to allow a new oversight agency. Meanwhile, certain developed countries were opposed, partly guided by a US/UK policy of 'no new agencies, no new money'. Caught between these two opposing sets of objections, negotiators were on the verge of having no competent sustainable development agency established. In the last month before the Summit, following the final preparatory meeting, the suggestion made by Ambassador Razali won the day.

Ambassador Razali and his supporters argued that there was a need to continue the discussion on means of implementation, which were not completed at Rio, and for a place to ensure the integration of the plethora of outcomes from the Summit. Their arguments won the day.

Agenda 21 explained it:

> In order to ensure the effective follow up of the conference, as well as to enhance international co-operation and rationalize the inter-governmental decision-making capacity for the integration of environment and development issues and to examine the progress of the implementation of Agenda 21 at the national, regional and international level, a high-level Commission on Sustainable Development should be established.
>
> (Agenda 21, 1992: 38.11)

On the other hand, the results of Rio weakened UNEP, partly because it had not very actively engaged in the preparatory process. UNEP had originally wanted to serve as the secretariat for the Summit, but had lost out to a newly established independent secretariat which would be able to draw from all UN bodies and beyond. This ensured an active involvement of the UN system as a whole, but did not put UNEP at the centre of the process.

The roles of the CSD and UNEP became increasingly blurred after Rio. During the first meeting of environment ministers following the Summit, at Brocket Hall in the UK, the chair's summary said:

> The high profile of the CSD whilst welcome has to some extent deflected political attention from UNEP. We agree that UNEP should be reaffirmed as an international focus for the environment.
>
> (Lindner, 1997: 6)

This became a recurring theme in the 1990s, until the 2000 meeting of Ministers in Malmö, which re-established and articulated a clear leadership role for UNEP.

Emergence of stakeholder democracy

The Earth Summit was an explosion of stakeholder involvement, similar in some ways to the mass uprising in Paris in 1968. But this time the stakeholders were creatively non-confrontational, and the authorities were not only listening but taking some of their ideas into international agreements.

Some of the key NGOs, including Greenpeace, Friends of the Earth and Third World Network, had at the final prepcom in April, published a ten-point plan to as they said 'save the Earth Summit'. Many of the issues had been discussed earlier, such as regulation of the multinational corporations, conserving the forests and changing the consumption patterns. Virtually none were picked up. This was because by April much of the negotiations were already locked into particular narratives between governments.

In the formal process, one of the significant outcomes was the nine chapters that, for the first time in a UN document, identified the roles and responsibilities of stakeholders. The UN had always bunched them together under the term 'non-governmental organizations'. Agenda 21 called them 'major groups'. Rio disaggregated them – acknowledging that each, in essence, plays a fundamentally different role in societies and nations. At the time, virtually no one could see how the 'major groups' approach would change the dynamics of the entire area of sustainable development governance.

The nine categories given a special role by Agenda 21 were: Business and Industry, Youth and Children, Local Authorities, Trade Unions and Workers, Non-Governmental Organizations, Women, Indigenous People, Farmers, and Scientists and the Technological Community. Clearly, other possible stakeholders could have been added. (Some of those since suggested include the education community, parliamentarians, the faith community, senior citizens and regional government.) But the representatives of the nine categories that Agenda 21 specified have shown themselves committed to being engaged in the follow-up to Rio in a way that couldn't have been imagined at the time. The UN Non-Government Liaison Service described this third section of Agenda 21 as:

the most extensive and formalized recognition in a UN Document of the potential and actual contributions of NGOs and other independent sectors.

(UN NGLS, 1997)

To facilitate the involvement and investment of those groups in Rio, the UN and the Brazilian government provided a huge open space for the first-ever major parallel meeting of civil society during an official summit. Coordinated by the NGO 'Centre for Our Common Future', and sited in beachside Flamenco Park, the Global Forum featured teach-ins, lectures, discussion groups, workshops and conferences on every possible issue concerning sustainable development.

The combination of intense political commitment and extensive personal interaction of individuals from every background and virtually every country produced an almost circus-like atmosphere where substantive issues of the greatest complexity were debated, decided and drafted, while street protests and theatre performances carried on only metres away. The Global Forum added up to an international environmental graduate seminar and cultural festival – inspiring a generation of policy activists and advocates to commit to a lifetime of action.

Simultaneously, there was an intense schedule of drafting sessions by NGOs, who were producing a series of 'NGO Treaties'. The International NGO Forum (INGOF) had been established during the UNCED preparations to facilitate policy dialogue among NGOs.[1] Crammed into tents that often reached steaming temperatures, hundreds of NGO officials, union organizers, indigenous leaders, academics, members of women's organizations, students and local individuals negotiated a multitude of issues in exquisite detail. By the time they were finished, they had produced 46 treaties dealing with all the subjects of the official Summit plus others that were not on the governmental agenda, such as energy and military issues. An early version of the Earth Charter was endorsed. The People's Earth Declaration opening the treaties said:

> We, the participants of the International NGO Forum at the Global Forum '92, have met in Rio de Janeiro as citizens of planet earth to share our concerns, our dreams and our plans for creating a new future for our world. We emerge from these deliberations with a profound sense that in the richness of our diversity, we share a common vision of a human society grounded in the values of simplicity, love, peace and reverence for life. We now go forth in solidarity to mobilize the moral and human resources of all nations in a unified social movement committed to the realization of this vision.
>
> (International NGO Forum, 1992a)

It is important to recognize that the organizations participating in the NGO Forum were trying to create a set of common commitments for action by NGOs, not providing an alternative to the official agreements. They saw themselves as innovators trying to help build a new sustainable society. As such, they created a model that, in the late 1990s, would be taken up under Brazilian civil society leadership and known as 'social forums'.

A simultaneously substantive and deeply symbolic ingredient of governmental and non-governmental deliberations was the involvement of representatives of the world's indigenous peoples. Leaders of indigenous populations travelled from every region of the world, and participated at a Sacred Earth Gathering of Spiritual Leaders, prior to the start of the official Rio conference.

The Declaration of the Sacred Earth Gathering called for a change in all human beings' values and behavior. Such sentiments were frequently integrated into the text and statements for the official conference.

Secretary-General Strong tried to capture this in his opening statement:

> We reinstate in our lives the ethic of love and respect for the Earth which traditional peoples have retained as central to their value systems. This must be accompanied by a revitalization of the values common to all of our principal religious and philosophical traditions. Caring, sharing, co-operation with and love of each other must no longer be seen as pious ideals, divorced from reality,

but rather as the indispensable basis for the new realities on which our survival and well-being must be premised.

(Strong, 1992)

Additional events at or parallel to the Global Forum included an industry and scientific conference, and exhibits portraying activities already being implemented to support sustainable development. EcoTech '92 had been a similar event earlier in the year, in São Paulo, Brazil.

Substantively most significant was the inclusion of civil society representatives inside the official negotiating process. UN negotiating protocol traditionally called for only the most insubstantial meetings (ironically called 'formal' sessions) to be open to non-governmental representatives. But the UNCED secretariat had made a consistent effort to open its preparatory meetings to those delegates.

Even in this, however, there was not smooth progress. When negotiations began at Rio, many of the negotiating sessions were closed to stakeholders, causing a rift between stakeholder organizations and governments. A strategy emerged to get around the problem by including stakeholder representatives on certain government delegations. Originally started by the UK, it had been taken up by many governments in order to keep the negotiations open. Approximately 150 stakeholders were recorded on governmental delegations at Rio. The strategy allowed stakeholders coalitions to stay up to speed on what was happening in the negotiating rooms and to refocus their lobbying accordingly. It also laid the foundation to open up most meetings in the various sustainable development processes to stakeholders as the 1990s progressed.

Chip Lindner the co-organizer of the Global Forum said:

Given the problems that confront us as a community of nations and peoples, we are now more than ever bound together by a common destiny. And solution to those problems will have to be found both nationally and internationally. That means that international institutions and national governments must become increasingly more accountable and responsive to the views and expectations of the world's peoples as a whole. Indeed, it means that as we approach the next century we must move even further in the direction of global democracy.

(Lindner, 1993: viii)

The building of public awareness and political momentum

Surprisingly, often the least recalled, but on some levels perhaps most significant, the Earth Summit was a massive success at gaining the attention of the world's public via global mass media.

For two weeks major news organizations in both developed and developing countries ran daily stories and special sections on the Summit, as journalists followed the background of the issues, the actions of their heads of state, and the colourful 'festival' of civil society representatives – all against a backdrop of Brazil's most spectacular beaches, most impoverished favellas and most opulent estates.

More than 10,000 journalists covered the government negotiations and the parallel civil society events, from virtually every country.

The media attention vastly raised awareness of environmental and development issues among the international public. It motivated a generation of educators around the world to add global ecological issues to their course curricula. And it prompted writers, celebrities and politicians to join the bandwagon and build momentum in support of global and local action.

Finally

The real measure of success, just as at Stockholm, would be what happened after Rio – what would government leaders, and stakeholders and citizens do when they got back to their countries, to their organizations, to their everyday preoccupations?

The Summit, it was hoped, would build the foundation to ensure that those decisions taken at the global level could be translated into action at national and local levels. It seemed that a new world order might be possible as nations moved into the twenty-first century, recognizing that everything depended upon uniting forces in a new global partnership which understood and respected the reality that all people had to live together on only one Earth.

In the 30 years prior to the Summit, the gap between the income of the 20 per cent of the world's population living in the developed countries and the 20 percent living in the poorest countries had grown from a ratio of 30:1 (in 1960), to 60:1 (in 1990).

As delegates streamed away from Rio, there were mixed views on what had been accomplished. Interestingly, most governments and NGOs presented a highly positive analysis.

> [the Summit] was to prove to be the most comprehensive social and economic agenda ever set before the UN in its 47-year history.
>
> (McCoy and McCully, 1993: 10)

> Northern officials and especially northern NGOs have become much more sensitized to the development needs and perspectives of the South ... Many environmental groups which in the past focused only on saving plant and animal life have come to a new understanding that resolving environmental problems requires tackling North–South and rich–poor inequities at the same time.
>
> (Khor, 1992: 8)

But perhaps the true reality was closest to being expressed by Secretary-General Strong in his closing remarks to the Earth Summit plenary:

> Finally, the remainder of this decade ... must be a time of transition which will truly move us on to the pathway to a new economy. The President

of one of the great corporations of our world told the Preparatory Committee in an informal session at its last meeting in New York that the present economic system is simply not adequate. This doesn't mean it needs to be scrapped, but it needs to be radically revised to bring it into tune with eco-realities. We need to move to a real economic system. ... The elimination of poverty has come through here as an important objective. But perhaps we haven't really committed to making this a central objective for the whole world community as we move into the 21st century. The New World Order, Mr. President, must unite us all in a global partnership which, of course, has to respect national sovereignty as a basic tenet, but must also recognize the transcending sovereignty of nature, of our only one Earth. The carrying capacity of our Earth can only sustain present and future generations if it is matched by the caring capacity of its people and its leaders. We must bring our species under control, for our own survival, for that of all life on our precious planet. Thanks to you, Mr. President, Distinguished Delegates, we now have a unique opportunity to do this. We have a basis for doing it in the decisions you have taken. We have the responsibility to start this road now. Our experience in Rio has been as historic and exhilarating as the road that brought us here. The road from Rio will be long, exciting, challenging. It will open a whole new era of promise and opportunity for our species if we change direction; but only if we start now. Mr. President, Distinguished Delegates, I think you all will agree that we must change the course that we have been on. That's why we're here.

(Strong, 1992)

The promise was there. But perhaps even more significant, and certainly of greater disappointment to delegates from developing countries, was the inability of UNCED to muster the financial commitments necessary to support all of Agenda 21.

Note

1   The Alternative Treaties were facilitated by the International Non-Governmental Forum (INGOF) Centre for Our Common Future Brundtland Bulletin No.16 op. cit. July 1992 p.7.

Chapter 4

Implementing Rio

> The process of change seems so gradual we have trouble recognizing it. From day to day, the lives of most of us seem not to change all that much. It is only when we lift our gaze beyond the next few days or years that we see the truth.
>
> (Gore, 1993)

One of the primary institutional outcomes of the Earth Summit was the establishment of the United Nations Commission on Sustainable Development (CSD). As the Stockholm Conference had added UNEP to the UN organizational family to deal with the environment, Rio added the CSD to attempt to integrate economic, social and environmental issues.

While the major international conventions launched at Rio had higher visibility and potentially stronger mandates to address their specific issues, the commission was given responsibility to perform the broadest task — to help coordinate the implementation of Agenda 21's sprawling constellation of efforts by governments, UN agencies and civil society to integrate social, economic and environmental policies.

In some ways, the CSD was set up as the successor organization to the Earth Summit itself — a permanent forum to collect attractive ideas, integrate them into coherent theory, institute those into manageable policy, and help implement that into effective practice.

It was an extremely broad set of responsibilities. And one of the immediate questions was whether the CSD had been given the institutional tools and resources to carry it out.

The CSD became the focus of high-level government involvement, mostly from environment ministries. This was going to be one of the main problems from the start. Overnight, the CSD became the main intergovernmental 'club' for environment ministries and environmental NGOs. And this was going to become one of its problems.

At Rio, environment and development ministries had come together to codify the fundamental principles of sustainable development, in close coordination with each country's foreign ministry and, frequently, trade and finance offices. But when the venue shifted to the annual CSD sessions in New York, representation from both the development and other ministries started to slip away – both because those agencies had other high profile processes to follow, and because it became increasingly apparent that there would be little new funding provided by developed countries for those development ministries to work with.

The creation of the Commission on Sustainable Development (CSD)

The CSD was established in 1993 as a functioning commission of the United Nations Economic and Social Council (ECOSOC). Agenda 21 had provided for the creation of the commission:

> In order to ensure the effective follow up of the conference, as well as to enhance international co-operation and rationalize the inter-governmental decision-making capacity for the integration of environment and development issues and to examine the progress of the implementation of Agenda 21 at the national, regional and international levels, a high-level Commission on Sustainable Development should be established.
>
> (United Nations, 1992a: Chapter 38.11)

As mentioned in the previous chapter, there had been considerable initial opposition from many Northern governments to the creation of the CSD.

However acting on the recommendations of the Earth Summit, the 1992 General Assembly debated setting up the commission and resolved that:

- The Economic and Social Council has been requested to establish a high-level Commission as a functional council body.
- Representatives of 53 states have been elected by the council for up to three-year terms.
- The Commission will meet once a year for two or three weeks. It is a functional ECOSOC commission with a full-time secretariat based in New York. Care has been taken to ensure that the secretariat has a clear identity within the UN system.
- Relevant intergovernmental organizations and specialized agencies, including financial institutions, are invited to designate representatives to advise and assist the Commission, and also to serve as focal points for the members and secretariat of the Commission between sessions.

In creating the mandate for the Commission, governments recognized the important role that Major Groups would have to play in the realization of the targets and goals of Agenda 21. Unquestionably, the Commission on Sustainable Development gives the Major Groups the greatest involvement in its work of any UN Commission.

The CSD's mandate was:

- to monitor progress on the implementation of Agenda 21 and activities related to the integration of environmental and developmental goals by governments, NGOs and other UN bodies. To monitor progress towards the target of 0.7 per cent GNP from developed countries for overseas development aid;
- to review the adequacy of financing and the transfer of technologies as outlined in Agenda 21;
- to receive and analyse relevant information from competent NGOs in the context of Agenda 21 implementation;
- to enhance dialogue with NGOs, the independent sector, and other entities outside the UN system, within the UN framework; and
- to provide recommendations to the General Assembly through the Economic and Social Council (ECOSOC).

The CSD secretariat was initially located within the UN Department for Policy Co-ordination and Sustainable Development (DPCSD). The DPCSD was one of three departments on social and economic affairs within the UN system. In 1997 it merged into the Department for Economic and Social Affairs (DESA). In addition to its role in the follow-up to the Earth Summit, the DPCSD included secretariats that would deal with outcomes from the Copenhagen World Summit for Social Development and the Beijing Women's Conference, both held in 1995.

The Head of DPCSD was UN Under Secretary-General Nitin Desai, who had been Deputy Secretary-General to Maurice Strong for the summit in Rio, and prior to that the Senior Economic Adviser for the Brundtland Commission. His appointment was greatly supported by governments and stakeholders alike.

The CSD was set up with 53 members – one fewer than the Economic and Social Council, the body it reports to. The allocations of seats are: 13 from Africa, 11 from Asia, 6 from Eastern Europe, 10 seats Latin America and the Caribbean, and 13 from Western Europe and North America. This allows for a full, proportional representation of governments from every global region, while, it is hoped, keeping that to a manageable total in terms of negotiating dynamics, consultations by the secretariat, logistics of meeting space and document production, and budgeting for all of these.

Each country has a three-year term of office on the Commission. Countries are designated to fill those seats by their respective regional bloc.

CSD and its relationship with the rest of the UN system

One of the innovations devised to encourage the integration of UN bodies was the creation of the UN task manager system. The task managers met through the establishment in October 1993 of the Inter-Agency Committee on Sustainable

Development (IACSD). Here 20 different UN agencies and programmes were designated to take responsibility for the implementation of specific Agenda 21 chapters and to identify major policy issues relating to the follow-up to UNCED by the UN system. The task managers met through the IACSD, established in October 1993. The IACSD focused on:

- streamlining the existing interagency coordination machinery;
- allocating and sharing responsibilities for Agenda 21 implementation by the UN system;
- monitoring the new financial requirements of UN system organizations that relate to Agenda 21;
- assessing reporting requirements that are related to the implementation of Agenda 21 and making recommendations on streamlining.

(IACSD, 1993)

The IACSD reported at first to the Administrative Committee on Coordination (ACC), made up of heads of agencies and programmes and chaired by the UN Secretary-General. In 1998 this body was superseded by the UN Chief Executive Board. It was later given responsibility for the Programme of Action for Small Island Developing States and other UNCED follow-up. The IACSD met twice a year, chaired by the DESA Under Secretary-General Nitin Desai. The meetings of the IACSD often brought in outside stakeholders and experts to report on their activities or suggest actions.

Many of the recommendations made in Agenda 21 were incorporated into the work programmes of the relevant UN agencies and programmes, but (and it was a big 'but') 'without additional funding' in most cases. As elsewhere at that time, development funding was simply not available. It is impossible to know how much more might have been accomplished with just a small amount of additional funding to support implementation within the UN system. But what is clear is that, working with little, a team of dedicated individuals who were committed to advancing the outcomes of Rio – led by an intense and creative director – took on significant personal strain and achieved an imperative institution-building task.

As the CSD collected the directives of governments and reported them, vertically, to ECOSOC and then the UN General Assembly, the IACSD ensured a horizontal linking within the UN system; the attempt to put in place an effective mechanism within the UN Division on Sustainable Development (DSD) was now established. What would governments do to mirror this?

This approach produced some interesting by-products: agencies and programmes were now answerable to a forum – the CSD – in addition to their governing body. The result of this peer group review was seen at the 1995 CSD session, when the Food and Agriculture Organization (FAO), the existing task manager for forests, was not given the role as secretariat for the newly established Inter-Governmental Panel on Forests, which was located instead in the UN secretariat.

Box 4.1 United Nations Commission on Sustainable Development Chairs 1993–1997

- CSD1: Razali Ismail, Malaysia, Ambassador to the UN
- CSD2: Klaus Töpfer, Germany, Minister of Environment, Nature Conservation and Nuclear Safety
- CSD3: Henrique Cavalcanti, Brazil, Minister of Environment and the Amazon
- CSD4: Rumen Gechev, Bulgaria, Deputy Prime Minister and Minister of Economy
- CSD5: Mostafa Tolba, Egypt, Former Executive Director of UNEP

Annual meetings of the UN Commission on Sustainable Development

First Session of the Commission on Sustainable Development – 1993

The CSD's opening session was timed to begin precisely one year after the climax of the Rio Summit. This symbolism was not lost on US Vice-President Al Gore. He had attended the Rio Summit as the leader of a delegation of US senators, with an independently earned reputation as an environmentalist and author of a recently published book, *Earth in the Balance* (Gore, 1992). The 12 months since had seen him surprisingly selected by Arkansas Governor Bill Clinton to be his vice-presidential running mate.

As he now appeared as the welcoming speaker at the CSD's opening session, Gore called for each country to make a serious commitment to change, and summed up the potential of the Commission:

> The role of this Commission is primarily catalytic – it can focus attention on issues of common interest. It can serve as a forum for raising ideas and plans. It can help resolve issues that arise as nations precede in their sustainable development agendas. It can monitor progress. It can help shift the multilateral financial institutions and bilateral assistance efforts towards a sustainable development agenda.
>
> (Gore, 1993)

There were other optimistic signs to emerge from the session. Delegates and ministers recognized that impromptu financial initiatives were very unlikely at the time. Consequently, they began to put in place the means by which money could be used in future should circumstances change. A pragmatic cooperation characterized much of the first meeting.

Chairing the first CSD meeting was Ambassador Razali of Malaysia, who had been the coordinator of institutional issues in the Rio Preparatory Meeting. His leadership – in particular in opening up the new commission to the involvement of stakeholders – changed the approach to basic governance practices within, and outside of, the entire UN system.

He helped to develop unprecedented access and involvement for stakeholders, principally through ad-hoc arrangements that led to the inclusion of verbal statements by stakeholders in 'informal' sessions, as well as in plenaries and at strategic points throughout the debates. Government delegates had been paying lip-service to the pivotal role that stakeholders play in promoting sustainability and the need to provide access to the intergovernmental process, but Razali's innovations made that access a reality.

The 1993 CSD agreed a three-year agenda (from 1994–1996) through which the Commission would review about a third of Agenda 21 each year. This would be followed in 1997 by a General Assembly review of the whole of Agenda 21.

Major Groups

The Major Groups had hundreds of representatives of NGOs and other civil society sectors, from every sectorial area and every global region. Crowded in to follow the first CSD session, it quickly became evident that they needed some form of coordinating structure, if only to help them each know what conference room meeting to attend. During the 1993 session, the predecessor organization to Stakeholder Forum conducted a survey among NGOs on what services might be needed between CSD meetings to help NGOs to prepare for the next meeting. On the evening of the closing day of that first CSD, its findings were presented to a large meeting of NGOs inside the plenary conference room. Those attending agreed to the need for a coordinating mechanism.

With the help of the UN Non-Governmental Liaison Service (NGLS), a series of regional telephone conferences were held in the autumn of 1993. They resulted in the establishment of an ad-hoc committee that met in Copenhagen in December 1993 at an NGO conference entitled 'Down to Earth – Between the Summits'. This conference brought together NGOs who had been involved in Rio with those involved with the 1995 World Summit for Social Development and the World Conference on Women. The ad-hoc committee agreed to establish an umbrella coordinating committee, and to employ the director of the US Citizens Network for Sustainability, Michael McCoy to draw up a fundraising proposal for an office to coordinate NGO preparations for the CSD. A small grant was subsequently provided by the Canadian Council for International Cooperation (CCIC).

Finance

The main sticking-point during the two-week meeting was the continuation of the debate on financing the commitments. Developing countries reminded their developed-country colleagues that the bargain in Rio had been that in return for agreeing to hold back on cheaper but environmentally destructive development, the developing countries would be provided with additional economic support. Yet virtually no new resources would appear that year. The failure was blamed on the continued world recession. Little did participants realize how bad it would get. By the end of 1993, ODA had fallen from US$60 to US$56 billion (OECD).[1]

Second Session of the Commission on Sustainable Development (1994)

The chair of the CSD second session was Dr Klaus Töpfer, then German Environment Minister. His selection demonstrated the high-level of political interest of national officials in the work of the Commission. It was an attempt to make the CSD not just another 'talking shop'.

Dr Töpfer had been an important minister in the Rio conference. The failure of the negotiations there for a Convention on Forests had been a real frustration for some governments, who wanted to recover the issue; this CSD offered that chance.

Dr Töpfer asked delegates and NGOs alike to 'give the issue air' – in other words, to allow an open discussion on the issue. The debate that followed succeeded in setting firm foundations for the creation of the CSD Inter-Governmental Panel on Forests, in 1995. The model of the Intergovernmental Panel on Climate Change provided a global strategy to address sustainable forest management.

Other issues

There were other significant outcomes from this session of the CSD, among which the greater cooperation between the Bretton Woods institutions, General Agreement on Tariffs and Trade (GATT), the World Trade Organization (WTO), the CSD and with major groups.

CSD recommended that governments should seek a legally binding status for the Prior Informed Consent Procedure that allowed for every country to be informed of any shipment of toxic wastes or products within its borders. It called for OECD countries to ban exports of listed or dangerous substances to developing countries. This ultimately became the Rotterdam Convention on the international importation of chemicals.

The development of a set of indicators for the relevant chapters of Agenda 21 was thought to be an important outcome for the future monitoring of implementation and to help provide information to governments for their decision making.

It had been standard practice that all UN commissions elected their chairs at the beginning of an annual session, and that those chairs would then work to the beginning of the next year's meeting. This was the model at the early meetings of the CSD. Clearly this made practical sense, as the commission meeting that a chair was leading had been set up by his predecessor. This is one of the issues that the second chair of the UNCSD took up to change, and after a number of years this was changed and then copied by other UN commissions.

National reporting

During the Rio process there had been considerable discussion of national reporting on the state of each country's challenges and achievements to the future CSD. Initially, developed countries had supported reports by each country that were compulsory and that would be peer group reviewed by other nations.

The compromise agreed to for the CSD was that five countries, one from each UN region, would offer their national reports voluntarily for review by other countries and stakeholders. This happened over a two-day period for each of the first three policy years of the CSD. While it turned out to be a very useful way for governments to share experiences and learn from each other, the voluntary nature of the process and the meagre number of nations participating left the Commission far from its assigned role of comprehensive international monitoring.

The Major Groups

At the CSD's second meeting, NGO delegates established an NGO coordinating structure with elected co-chairs from developed and developing countries and an elected representative committee. The NGO Steering Committee had two functions:

* to facilitate and manage the involvement of NGOs in the CSD; and
* to help facilitate access to the process for all nine Major Groups.

It therefore included representatives of all the Major Groups as members, and a structure that contained NGO issue caucuses and regional caucuses. Each caucus elected two co-chairs who had to be balanced between the developed and developing countries, and between male and female. The Committee was far broader in its definition of regions or sub-regions than the intergovernmental structures, and the eventual make-up of the Committee had around 70 per cent of its regions from developing countries. At the time, it represented the most advanced form of stakeholder governance body around an intergovernmental process, and bridged the splits that had been so evident in the run-up to Rio.

Following the success of the 1994 CSD session there was an explosion of more than 100 'intersessional meetings', government-sponsored or NGO meetings, feeding into the 1995 CSD – many of them on forest related issues.

Finance

The 1994 CSD session recognized that the overall financing of Agenda 21 and sustainable development fell significantly short of expectations and requirements.

This was going to become a recurring and increasingly contentious theme in future sessions. By the end of 1994, ODA had risen slightly, to US$ 59 billion (OECD).

Third Session of the Commission on Sustainable Development (1995)

The third session of the CSD capped a period of strong support for the commission. The new chair was Brazil's Minister of Environment and the Amazon, Henrique Cavalcanti. That year's CSD saw its largest number of national reports produced, 53, with virtually every country participating. Many of these included inputs from their nation's stakeholders. Ten countries made presentations on their national reports.

Some governments began setting up national councils for sustainable development, among them the US, who created the President's Council on Sustainable Development. More than 50 ministers and secretaries of state attended the third CSD session, including ministers of the environment, finance, planning, development cooperation, forestry, agriculture, labour and infrastructure.

Another significant feature was the growing collaboration of the entire family of UN institutions, both during the preparatory phase and the session itself. A pervading atmosphere of optimism and expectation was apparent in the statement made by Under Secretary-General Nitin Desai on the opening day:

> We have to appreciate that the CSD is something unique among ECOSOC bodies. It has not only managed to mobilize the interest and active involvement of the UN system, including the development banks, but it has captured the attention of non-governmental groups and the public at large. The Commission has gotten this response because of the urgency of its subject matter and the open and transparent way it has conducted its business. The effort to be inclusive rather than exclusive has generated support for the Commission and commitment to its work programme.
>
> (Desai, 1995: 2)

The greatest success of the 1995 CSD was undoubtedly the establishment of an Inter-Governmental Panel on Forests (IPF). Its remit was to recommend whether there should be a new convention on forests, and report to the Special Session in 1997. It was agreed that the panel should consider:

- implementation of UNCSD decisions related to forests at the national and international level;
- international cooperation in financial assistance and technology transfer;
- scientific research, forest assessment and development of criteria and indicators for sustainable forest management;
- trade and environment relating to forest products and services; and
- international organizations and multilateral institutions and instruments including appropriate legal mechanisms.

(UNCSD, 1995)

To support the IPF and to assist governments to implement the proposals for action, an informal, high-level Interagency Task Force on Forests (ITFF) was created, made up of eight international organizations.

> The setting up of an Intergovernmental Panel on Forests was unanimously supported, and seen as a real achievement demonstrating the level of credibility attained by the CSD in fulfilling one of the main decisions reached at UNCSD. [However], much disappointment was expressed with regard to the need for new and additional resources in terms of ODA, which has declined both in absolute terms and as a percentage of GNP.
>
> (Cavalcanti, 1995: 1)

There were three other significant, but less well-known outcomes. A programme of work was established on consumption and production patterns, a timetable was agreed for the formulation of sustainable development indicators, and initiatives were encouraged to phase out lead in petrol, at national and international levels. Lead was outlawed within five years in Europe and by the year 2000 in the US. By 2010, only eleven countries still permitted lead in gasoline.

The Major Groups – Local Authorities

Over 300 stakeholders attended this CSD. The previous year, the International Council for Local Environmental Initiatives (ICLEI) had secured reference in the formal texts to the desirability of arranging an organized reporting process on the role of local authorities in implementing Agenda 21. During the 1995 session, negotiations were successfully concluded for the preparation and circulation of a collection of local Agenda 21 case studies from all regions of the world, the mounting of a local Agenda 21 exhibition during the CSD session, and the presentation of a local government case study from each region of the world on 18 April 1995. This 'Day of Local Authorities' was seen as an effective way to highlight concrete actions at the local level by specific Major Groups. This event has been continued at subsequent sessions.

Finance

One of the continuing areas of concern remained the financing of sustainable development, especially to support national efforts in developing countries and economies in transition. By the end of 1995, ODA was still around US$59 billion (OECD).

In the high-level segment of the CSD the UK Minister John Gummer underlined the importance of the CSD when he said:

> This is a ministerially driven organization determined to see that the people who actually make the decisions nationally play their proper part internationally.
>
> (Gummer, 1995: 62)

Fourth Session of the Commission on Sustainable Development (1996)

The selection of Rumen Gechev, Deputy Prime Minister and Minister of Economy of Bulgaria, as chair of CSD 1996 was an attempt to arrest the movement of the CSD becoming an environment minister's forum. Unfortunately, this coincided with that country's serious financial crisis (1995–1997), caused in part by the delay in the formation of a market economy. The chair was therefore often focusing elsewhere than the CSD.

The 1996 CSD was in many ways a disappointment, particularly after the success of 1995. Since the Inter-Governmental Panel had helped to focus work on forests in one forum, there was an effort for the CSD to do the same in 1996 for management of the marine environment. This was in part led by the UK Prime Minister's Advisory Panel on Sustainable Development, whose chair called for an Inter-Governmental Panel on Oceans as a way of advancing the issue. It was taken up by the UK Government, but the CSD did not agree to set up an equivalent structure.

With the General Assembly review of Agenda 21 approaching, much of the CSD meeting was spent considering how the five-year review should be conducted, and what the priorities for consideration should be.

The Commission failed to agree recommendations to the International Maritime Organization on the need for negotiation of a legally binding, global agreement on oil platform discharges and on 'user pays' fees for the upkeep of the straits used for international shipping.

In the area of changing consumption and production patterns, the CSD reviewed the work programme but could only conclude that although eco-efficiency was a promising strategy for policy development, it was not a substitute for changes in the unsustainable life-styles of consumers. The Commission urged governments to look at their procurement policies to see if they could be 'greened'.

> The Commission on Sustainable Development was created in an effort to deal politically with the entire emerging agenda from the Rio Conference. Our consumption patterns, one of the root causes of unsustainability, have at last entered the UN agenda. Will the UN be able to act on the problem? We would fail abysmally should the answer be no. What is required is a much more powerful voice on the part of the CSD.
>
> (Brundtland, 1996: 30)

Achievements

Following up on the Global Plan of Action negotiated at a Washington, DC meeting on protecting the marine environment from land-based activities, in November 1995, a draft decision on institutional arrangements was agreed for consideration by the UN General Assembly. The Plan of Action gave UNEP a new and enhanced role in this field.

Continuing its 1995 discussion on indicators, the Commission urged governments to pilot the 126 indicators developed by the CSD in conjunction with governments, UN agencies and major groups.

The Major Groups

After the success of the Day of Local Authorities, the 1996 CSD dedicated a 'Day of the Workplace' bringing together industry and trade unions. A series of impressive examples of how those major groups are involved in implementing Agenda 21 were presented.

> In its interventions at the CSD, the ICFTU has consistently reiterated that eco-auditing measures could be implemented quite simply and has proposed that it begin focusing on four key areas of change within each workplace: energy conservation, resource management, pollution control and waste management.
>
> (Royer, 1996: 26)

Just as governments were preparing for the five-year review, stakeholders were also seeking to define the agenda for 1997. The CSD NGO Steering Committee started a call for input to what would become known as 'Towards Earth Summit II'.

Finance

By 1996 ODA was falling again, now down to just under US$57 billion (OECD). In the discussion on finance there was still a call for increasing aid but also for making it more effective. The Danish Environment Minister, representing one of the few countries which had reached 0.7 per cent GDP threw down the gauntlet to other developed countries:

> In my own country, Denmark we are currently moving our ODA from the 1 per cent GDP we reached in 1992 towards our new target of 1.5 per cent to be reached by 2002.
>
> (Auken, 1996: 30)

Note

1 OECD figures for ODA 1993–1996 are taken from OECD Aid Statistics, available online at http://www.oecd.org/dac/stats/data

Chapter 5

A wake-up call

The United Nations General Assembly Special Session (UNGASS) to Review and Appraise the Implementation of Agenda 21 crept up on most people without the kind of preparation that could have worked to make it a more significant event. Even its acronym – UNGASS – seemed unconsidered and unplanned. One NGO observer joked that it must be either a subliminal effort to bring the issue of carbon dioxide emissions onto the review agenda, or an unwitting prediction of the session's results.

Unlike the summit in Rio, where an independent secretariat worked for two and a half years on planning the event, the General Assembly review utilized the already-assigned secretariat of the CSD. This meant that staff only became free to work on the review at the conclusion of the CSD meeting, in May 1996.

The late schedule also had an impact on governments' approach to the review. The first consultation for governments – a retreat hosted by Oxford University's Green College and Stakeholder Forum – wasn't until June 1996. It included papers that had been commissioned to focus on critical issues, including one on future risks.

The CSD NGO Steering Committee – the broad international coalition of NGOs active at the CSD – produced a carefully-written document as the NGO input to the negotiations, titled 'Towards Earth Summit II'. The product of extensive internal negotiations, it covered virtually every issue on the table at the Special Session – and a few others that were not. It also included a list of top ten priority actions that were required of governments, which proved an effective lobbying tool.

The preparatory process for the special two-week intersessional in late February, the three-week CSD in April and a final prepcom week in mid-June, just before the Special Session (23–27 June). In total, 35 days of negotiations, that didn't compare to the 75 days for Rio itself, but wasn't meant to. The Special Session was meant to be only a review, a chance to correct the direction governments were going in and to address any new areas that might now require attention. It was clear going into the first planning meeting that forests would be on the agenda, as well as the 'missing' energy chapter of Agenda 21. But the most important issue for virtually all developing country governments was the disappearing ODA.

Box 5.1 **The CSD NGO Steering Committee published the NGO input through a document called 'Towards Earth Summit II'**

Its 'top ten' list of priority actions were:

1. Climate: A legally binding agreement reducing CO_2 emissions to 20 per cent below 1990 levels by 2000.
2. Toxic chemicals: A legally binding agreement on persistent organic pollutants (POPs) by 2000.
3. Forests: The continuation of intergovernmental dialogue on forests under the CSD and the establishment of a network of ecologically representative and socially appropriate forest protected areas (covering at least 10 per cent of the world's forests) by 2000.
4. Freshwater: Initiating negotiations on a global agreement on freshwater by 1998.
5. Military weapons: A global legally binding agreement on anti-personnel land mines by 1999.
6. Finance: Fulfilment of the ODA target of 0.7 per cent by 2002, a 50 per cent increase in GEF replenishment, and a 20 per cent debt reduction of the heavily indebted least-developed countries by the end of 1998.
7. Trade: Adopting an international agreement that multilateral environmental agreements (MEAs) shall not be bound by WTO requirements by 1998.
8. Corporate accountability: Establishing a sub-commission of the CSD to ensure significantly greater accountability of business and industry.
9. Indigenous peoples: Establishing a Permanent Forum for Indigenous Peoples within the UN System.
10. NGO access: Ensuring that the rules of procedure for NGOs at UNGASS are similar to the UN CSD process.

(Dodds *et al.*, 1997: 48)

Planning Meeting 1 – The CSD Intersessional Working Group

At the start of the intersessional meeting in February 1997, two co-chairs were selected to guide the process – Celso Amorim, the Ambassador to the UN from Brazil, and Derek Osborn, the recently-retired director of the UK Department of Environment who was now chair of Stakeholder Forum.

The meeting decided to define the structure of the Special Session through four outcomes documents:

- a statement of commitment – to set out the political commitment and the challenges being faced;
- an assessment of progress reached after Rio – to describe the effects of globalization; the economic, social and environmental trends; the vital universally declining finance and ODA trends; the state of technology transfer, the present status of international environmental agreements, and the participation of Major Groups;
- a review of implementation in areas requiring urgent action;
- a survey of international institutional arrangements – to address coordination among intergovernmental organizations, and role and agenda of the CSD.

The first meeting's goal was to produce a draft negotiating text for governments to start work on. The chairs asked for governments and stakeholders to contribute ideas that reflected 'maximum creativity and receptivity' at that early stage.

Over the two weeks of the meeting the debate covered the whole of Agenda 21 but focused on six areas:

- combating poverty and the growing inequality, particularly in the countries where poverty and environmental degradation are most acute;
- stopping the decline of ODA that by the end of 1997 had fallen to under US$49 billion (OECD),[1] a decline of nearly 20 per cent since the Earth Summit;
- supplying freshwater and sanitation to the billions of people without each;
- achieving a clear global deal on climate change;
- establishing a process to promote sustainable management of world forests; and
- protecting the marine environment and reversing the decline of fish stocks.

One of the early breakthroughs for NGOs had been the General Assembly's decision to introduce multi-stakeholder dialogues during the Special Session's final preparatory meeting. The dialogues would highlight what each stakeholder group had achieved and what it felt governments should address as priorities. In line with this more inclusive approach, one of the group's suggestions – that each of the governments should have completed drafting their sustainable development strategies by 2002 – was picked up by Pakistan Ambassador Shakquat Kakakhel, who had chaired the Habitat II Conference on sustainable communities the previous year – still considered the most advanced example of stakeholders' involvement in any intergovernmental meeting.

At Habitat II, key stakeholders were literally allowed to have a seat at the table, and suggest text into the negotiations, and if a government took up the suggestion then the suggestion became incorporated into the negotiating draft. Ambassador Kakakhel had championed many of these innovations, and he went on to be Deputy Executive Director of UNEP a year later.

Rio+5 Forum

From 13–19 March 1997 the Earth Council convened a forum to assess the progress since Rio and provide a means for sustainable development practitioners to input to the Special Session. The event was attended by over 1000 stakeholders, heads of UN programmes and national councils for sustainable development.

The forum attempted to assemble a set of recommendations to submit to the CSD, and to gain backing for the Earth Charter. Neither was possible. The positioning of the forum between two formal preparatory sessions probably cost it the impact hoped for. Many NGOs had already presented their list of priorities in New York, and governments were now responding to the chair's text that had emerged from the Intersessional Working Group meeting. Perhaps the conflicts in timing were also an early indication of the problems that would be faced at the five-year review on a number of fronts.

Planning Meeting 2 – Commission on Sustainable Development

The new chair of the Commission on Sustainable Development was Mostafa Tolba of Egypt, the former Executive Director of UNEP, and a seasoned facilitator of intergovernmental negotiations.

For its first five years, the CSD had been fortunate to have the indefatigable Joke Waller-Hunter as director of its secretariat, the Division on Sustainable Development. At the opening of this extraordinary CSD session, she outlined three criteria by which to judge the success or failure of UNGASS. These were:

- Does the assessment reflect the urgency of the situation?
- Is the assessment followed by a unequivocal commitment to concrete action?
- Have partnerships been acknowledged, renewed and strengthened?

A number of ministers attended the first week of the CSD and they were able to instil some political direction to the negotiations. The European Union launched a series of initiatives on freshwater, energy, and on sustainable consumption and production. Significant improvements to the forward planning of the CSD itself were also agreed, enabling attention to be focused thereafter on a limited number of topics to be explored in depth rather than attempting to review progress on the whole of Agenda 21 every year.

The CSD meetings provided space for the precedent, setting half-day presentations by each of the nine Major Groups on their achievements and the challenges they were facing. Unfortunately these were scheduled parallel to negotiations on the primary Special Session document, so not many governments sent representatives and therefore the debate was poor. The chair did reveal, however, that stakeholders were successfully taking forward Agenda 21 in ways that governments were not.

The presentations provided good examples of activities in which Major Groups were taking the lead. The Local Governments group announced that there were now over 1800 local Agenda 21s, in 64 countries. Trade unions reported examples of achievements in such countries as the Philippines, where the San Miguel Corporation, an industry leader, had agreed to an environmental protection clause into the collective bargaining agreement with the Congress of Independent Organizations – Associated Labour Unions (CIO-ALU) to establish labour–management cooperation for environmental protection.

At the end of the first week, governments decided that they needed more time for direct negotiations, so the decision was taken to cancel the final week of the CSD, and instead hold a week of 'informal' meetings before the Special Session. Leaving the CSD, the major areas of disagreement were on forests, climate, and poverty eradication and financing support for developing countries.

The Special Session

Coincidentally the additional week of negotiations occurred just days after the 20 June adoption by the UN General Assembly of the Agenda for Development, a process that had started in 1992. Many developing countries were highly disappointed with the outcomes of this process, and a number brought their frustration and anger into the Rio+5 process over the next two weeks.

Finance

That frustration was particularly evident in the area of finance, which had already played central role in the disagreements during the preparatory process.

As their outgoing contributions of ODA continued to drop even although their economies started to boom, the major industrialized countries developed a mantra that its place would be taken by foreign direct investment (FDI), the money that private businesses from developed countries spent to buy businesses, build factories and employ workers in developing nations. Some developing countries were deeply sceptical of the probable amounts of corporate funds that would be invested – and whether such amounts would stay in-country to benefit their economies and people, or simply be extracted as profit and returned to the investors' home country.

Foreign direct investment *had* increased substantially over the previous ten years – from US$25 billion to over US$170 billion – but almost three-quarters of it had gone to only twelve countries and there was no evidence that any of it had gone to support the implementation of Agenda 21. Efforts were made by some of the more progressive donor countries to secure firm commitments to increase overall aid levels towards the long-established goal of 0.7 per cent of GDP. But in the end these efforts only served to advertise disagreement between the donor countries, and no significant progress on levels of aid could be achieved.

As one NGO at the Session observed:

> Finance debates at Earth Summit II were an unedifying dialogue of the deaf. Developed countries' repeated references to the importance of private

investment for development inflamed the South's legitimate frustration at declining aid budgets.

(R. Lake quoted in Osborn and Bigg, 1998: 20)

For a year a group of NGOs had been working on the idea of an Intergovernmental Panel on Finance, recognizing that there was going to be a collision on finance at UNGASS. The effort was led by Barbara Bramble of the US-based National Wildlife Federation (NWF), who had been one of the NGO leaders in Rio and was supported by the Royal Society for the Protection of Birds (RSPB) and Stakeholder Forum. It built on the work of the Japan/Malaysia expert group which had been looking at different options available for financing sustainable development. NGOs suggested the remit of such a finance panel would be to look at:

- ODA contributions – quantity and quality;
- domestic mobilization of resources;
- foreign direct investment; and
- new financial mechanisms.

Their proposal was championed by the Norwegians and, surprisingly, the United States. But because of its distrust of the promises of developed nations, G77 refused to accept the panel without a firm timetable on ODA and a more specific acknowledgement of the failures to deliver on promises from 1992 (OECD, n.d.).

Explaining the position of G77, Ambassador Mytusa Waldi Mangachi of Tanzania said:

> There is a need for political will that starts with the grassroots people of the North and works up to the politicians and political agencies to the international level: the G8, the World Bank, the IMF and the Paris Club. You know when the North deals with the South on financial issues they always have their 'clubs'. But we in the South always end up negotiating as individual countries. A panel on finance will not provide political will, only more bureaucracy.
>
> (W. Mangachi quoted in Osborn and Bigg, 1997: 55)

The panel idea remained one of only two issues that were not resolved at UNGASS and were sent to the General Assembly. The US made a commitment of US$100,000 to support creating a space for the panel ideas to be discussed. Although still defeated in the General Assembly, the process continued informally and eventually became the Financing for Development process, which in the early 2000s targeted least-developed countries, particularly in Africa, and was considered a significant success.

Another successful campaign by the NGOs had been to persuade the EU to change its position to support an international aviation tax which would be hypothecated to ODA for sustainable development. The suggestion remained in the text until Argentina opposed it.

Forests

Forests were another critical issue. A number of developed countries – particularly Canada and the EU, with Germany and the Netherlands as its leaders – had worked for UNGASS to call for the setting up of an intergovernmental negotiating committee (INC) to advance negotiations on a possible new forest convention. The US had changed its position since the Rio conference and now favoured it. But ironically, most NGOs had shifted positions also and were now opposed. The lead NGO arguing for a convention was the Environmental Investigation Agency (EIA) with support from some smaller NGOs. The larger NGOs, however, were worried that negotiating a convention now would weaken current standards and support commercial interests over conservation. They and others argued for continuing a process under the CSD. This was the eventual diplomatic fudge – a new body called the Intergovernmental Forum on Forests (IFF) would be established, and the issue on a convention would be revisited in 1999. However the debate did, due to EIA, put illegal logging on the agenda, and it became part of the environmental crime discourse over the coming years.

Climate change

The high-level segment of the UN General Assembly Special Session opened in late June with typical diplomatic fanfare – and a surprising number of appearances by heads of state of major nations. With the session artfully – or accidentally – scheduled to open only days after that year's summit of the G7 economic powers in Denver, in the western US, leaders of those nations found it convenient to travel to New York for the high-profile session.

A critically important climate change negotiation was rapidly approaching that December, in Kyoto (UNFCCC COP3), with industrialized nations under intense pressure to announce specific national reductions in their carbon dioxide emissions to below the level of 1992 (when the UNFCCC had been agreed). In New York, each of the major EU countries' leaders – Blair of the UK, Kohl of Germany, Chirac of France – had announced targets with significant reductions, as had Japan, Canada, the Netherlands and each of the Nordic nations.

With a Democrat in the US White House and the American economy booming, NGOs had hoped – and indeed many not well-grounded in US domestic politics had assumed – that the US would play a far more constructive role in committing to progress at climate negotiations than it had under the famously obstructionist President George H.W. Bush. But President Bill Clinton, despite positive words, was concerned about criticism from energy corporations and faced a virulent political opposition, and had been extremely limited in going beyond the actions of his predecessor.

So when he addressed the UNGASS plenary on the closing session, his words were listened to with intense fascination. He began, positively, with a well-known recitation of the statistics:

With four per cent of the world's population, we produce 20 per cent of its greenhouse gases. Frankly our record since Rio is not sufficient. We have been blessed with high growth for several years, but that has led to an increase in greenhouse gas emissions in spite of the adoption of new conservation practices. So we must do better, and we will.

(Clinton, 1997)

But then he announced his position for Kyoto. The US would stabilize, not reduce its emissions, at 1992 levels. President Clinton trumpeted it as a brave and virtuous action. Others were deeply disappointed. In public statements by the leaders and private communications by delegations, the EU and other industrialized countries made it clear that the US position was insufficient. President Clinton was reportedly stunned that the leaders he had just played host to could have 'turned' against him so vocally. In an exchange overheard in a General Assembly elevator, a high-ranking member of the US delegation attempted to berate one of his equivalent-ranking colleagues, only to be told, in effect, 'We've had enough of waiting for you and listening to words. It's time to act'.

Nevertheless, late on the last evening of the Special Session, responding to the pressure generated at the meeting, all countries agreed to adopt a form of words looking forward to further progress at Kyoto, indicating that many but not yet all countries intended to adopt significant targets and measures at the Kyoto meeting.

At a UN press conference that Friday, after waiting well into the early evening to hear the full US position, a panel of five highly respected environment and development NGOs, from all regions, added quotable ammunition to the attack on the American position.

Only a couple of days after major lobbying efforts by the fossil fuel industry which ran full page advertisements in the US media, Clinton was not prepared to deliver what is needed. Over the last few days the Special Session has reaffirmed that climate change is the most important environmental issue on the world stage, in reaffirming the need for 'significant reductions', Clinton can only postpone but not evade the question of targets and timelines if he wants to remain credible in international negotiations.

(Hare, 1997)

The NGOs made it clear that the United States – and President Clinton – would be held responsible by history if Kyoto failed. Six months later, after much international debating and an internal campaign by US NGOs, the US representative at the Kyoto protocol talks, Vice-President Al Gore, announced a new US position – reductions relative to 1992 levels of 2 per cent. Not that much of a difference, and a far cry from the levels that all would agree were eventually needed, but a significant shift that was significant – and just enough to allow the Kyoto protocol to go through to agreement.

While hardly the only factor, the Special Session may have played a significant or even critical role by airing some of the key issues on climate change in an early, public setting, and allowing leaders to say and hear what each really thought.

New issues

Although UNGASS was primarily meant to review implementation of the existing Agenda 21 issues, governments at the session also added to the agenda three new issues that were not in the agreement in 1992 – energy, transport and tourism.

Energy

Energy had been a serious failure in Rio, as objections from an unrelenting group of OPEC oil-producing countries – with quiet, and sometimes open, backing by the US and Australia – repeatedly blocked any attempt to complete a chapter on energy for Agenda 21. At the Special Session five years later, the EU led a determined effort to negotiate a compromise with the G77 that would resolve the fundamental sustainable development challenge of balancing the necessity of stabilizing overall energy consumption with increased 'access to energy' by billions of the world's poorest. With steady encouragement by NGOs, a compromise agreed at UNGASS set in motion an intersessional process on a review of energy that would culminate at the CSD in 2001, where it was planned that more substantial results would be agreed.

Transport

Traditionally considered a national issue, transport was increasingly revealing areas that impacted on sustainable development internationally as well as nationally. Among the issues being considered were setting standards for lead in petrol and a proposed aviation tax by the EU. Broader issues such as building integrated transport systems and government support for mass transit went to the core of practically addressing climate change, energy reduction, access to services for the poor, health issues and encouraging sustainable cities.

Sustainable tourism

While not widely thought of as an issue in Rio, by the time of the Special Session tourism was starting to gain traction. In preparing for the review of Agenda 21, the World Tourism Organization, UNEP and UNESCO held a Sustainable Tourism Conference in 1995 in Lanzarote, Spain, which produced the World Travel and Tourism Council (WTTC) Lanzarote Declaration on Sustainable Tourism. The following year an initiative by the Earth Council, WTTC and UNEP produced *Agenda 21 for the Travel and Tourism Industry: Towards Environmental Sustainable Development*. Governments at UNGASS recognized sustainable tourism as a significant issue and agreed to address it at the next meeting of the UN CSD, and to develop a work programme.

Conclusion

> One of the characterizations of the five years since Rio was that they have
> produced little more than 'Slightly Less Unsustainable Development
> Genuflecting to the Environment' or 'SLUDGE'. It was urged at the beginning
> of the review meeting in New York that what the world needs now is to move
> from 'SLUDGE' to 'Development Reconciling Environment And Material
> Success' or 'DREAMS'.
>
> (Osborn and Bigg, 1997: 3)

Despite the limited time for preparation, and serious obstacles resulting from the
increasing schism between developing and developed nations, the Special Session
produced significant progress on the Special Session issues. But in reality a great deal
of 'SLUDGE' remained; and to a large extent 'DREAMS' remained a dream.

There were some lessons from 1997 which would prove useful to take to the
Rio+10 Conference in 2002. These included the necessity for adequate preparatory
time for governments and stakeholders to input to the process in order to make
progress on new issues, and to identify and address the broad underlying political
dynamics that could threaten any agreement.

The lack of trust between developed and developing countries had grown
through the UNGASS process as it became even clearer that ODA was falling
drastically since the conference in Rio. Developing countries were facing the
possibility of a lost decade. Would UNGASS be a catalyst for ODA commitments,
and would they start going back up? Delegates recalled that in 1992 the UNCED
secretariat had estimated it would cost US$125 billion a year to implement Agenda
21, yet development aid was now at the lowest level since 1989, standing at just
under US$49 billion (OECD).

For the next five years (and the larger review in 2002) to be a success there would
literally need to be a new deal between developed and developing countries on
finance, technology transfer and trade to support sustainable development.

NGOs had low expectations from UNGASS but of their top ten issues they had
seen success for their positions on toxic chemicals, forests replenishment of GEF, and
NGO access. This amounted to a 3.5 score out of 10. By 2011, of those original 10,
close to 8 had been achieved – an example both of the importance of NGOs setting
agendas and building coalitions, and of realizing that in international policy, progress
is often measured in decades, not days.

Note

1   OECD figures for ODA 1993–1996 are taken from OECD Aid Statistics, available
    online at http://www.oecd.org/dac/stats/data

Chapter 6

Rebuilding momentum

The UN General Assembly Special Session (UNGASS) had been a real wake-up call for developed countries' heads of state. Many had been stunned by the unrelenting criticism by developing countries of the sincerity of their promises on overseas development aid (ODA), and by the threats of those countries to block all negotiations on the environmental issues. UNGASS had marked the low point in development aid. And it wasn't until the next year that ODA started going back up – to US$52 billion, which would be its general level until after the Millennium Development Summit in 2000.

Going into 1998, however, there was some increased optimism regarding ODA among the development community, particularly following the success of the Kyoto Protocol negotiations in December 1997. It now looked like climate might at last have gained the momentum that had initially been called for at the Toronto Conference nearly 10 years earlier.

Unfortunately, any such progress would have to be achieved without the US, whose domestic politics had left it increasingly hostile to any intergovernmental agreements or financial obligations (any, at least, that did not involve other governments' near-total acquiescence to Washington). This had become an increasingly recurring theme over the previous 20 years.

The US Senate made it resoundingly clear it would not ratify the Kyoto Protocol in a June 1997 vote of 95 to zero. Its resolution read:

> It is the sense of the Senate that:
> 1. the United States should not be a signatory to any protocol to, or other agreement regarding, the United Nations Framework Convention on Climate Change of 1992, at negotiations in Kyoto in December 1997, or thereafter, which would
>   • mandate new commitments to limit or reduce greenhouse gas emissions for the Annex I Parties, unless the protocol or other agreement also mandates new specific scheduled commitments to limit or reduce greenhouse gas emissions for Developing Country Parties within the same compliance period, or
>   • would result in serious harm to the economy of the United States; and

2. any such protocol or other agreement which would require the advice and consent of the Senate to ratification should be accompanied by a detailed explanation of any legislation or regulatory actions that may be required to implement the protocol or other agreement and should also be accompanied by an analysis of the detailed financial costs and other impacts on the economy of the United States which would be incurred by the implementation of the protocol or other agreement.

(US Congress, 1997)

Sixth Session of the Commission on Sustainable Development (1998)

This was the background against which Philippines Minister of the Environment, Dr Cielito Habito, took over as chair for the sixth session of the CSD. He was to preside over a commission trying to regain its momentum and introduce an innovative process that stakeholders had persuaded governments to accept at the Special Session – a two day interactive dialogue between civil society organizations and governments.

The dialogue would comprise four formal UN CSD sessions, each covering three hours – two full days of discussion. It would be scheduled at the beginning of the two-week CSD meeting, so as to enable governments to take serious note of stakeholder views prior to their negotiating of official agreements.

To help design this new approach, the UN Division on Sustainable Development (DSD) had hosted a retreat that included key stakeholders representing trade unions (the ICFTU), business (the WBCSD) and NGOs (the NGO CSD Steering Committee).

The consultation had been intense and productive and the outcome was a broadly inclusive dialogue format that produced one of the most spontaneous exchanges of opinion that the CSD had so far seen. It afforded stakeholder groups the opportunity to comprehensively analyze policies and to put governments under pressure to defend their positions. Also fascinating was that, for the first time, governments could put stakeholders under pressure to defend what they said.

Box 6.1 **United Nations Commission on Sustainable Development Chairs 1998–2002**

- CSD1998: Dr Cielito Habito, Philippines, Minister of the Environment
- CSD1999: Simon Upton, New Zealand, Minister of the Environment
- CSD2000: Juan Antonio Mayer, Colombia, Minister of the Environment
- CSD2001: Bedřich Moldan, Czech Republic, former Minister of the Environment
- WSSD2002: Emil Salim, Indonesia, former Minister of the Environment

1998 multi-stakeholder dialogue on industry

The 1998 dialogue sessions were organized to address the broad issue of 'Industry and Sustainable Development', and four subtopics:

- responsible entrepreneurship;
- corporate management tools;
- technology cooperation and assessment; and
- industry and freshwater.

Some governments were uncomfortable about giving away twelve hours of potential negotiating time, and about the enhanced role that stakeholders were being given. They questioned whether the discussions would produce any meaningful results, and whether the dialogues were worthy of appearing in such an august venue.

That first year's dialogues included participation by three of the CSD's nine civil society sectors – trade unions, business and NGOs. They were held in the UN's ECOSOC Chamber, whose soaring tapestries, curving delegate's tables and softer acoustics lend themselves to a more interactive debate.

The first day's session showed that intense interactions could occur not only between governments and stakeholders, but among stakeholders themselves. During the opening statements, representatives of the business group had described in calm, confident tones how voluntary guidelines established by industries would be sufficient to achieve safe products, environmentally friendly production and healthy working conditions. But in the afternoon session, apparently distressed by a newspaper article he had just read, one of the lead business representatives spoke in a much more agitated way.

> I think it is very important that we do this in a spirit of trust … We have come here as business in a spirit of trust and to have a trustful dialogue. I find to my enormous surprise that there was an article today by one of the participants here from Friends of the Earth and it says, 'It seems that multinational corporations regard regulations as fine to protect business, but if it protects the planet it is unnecessary and bureaucratic.'
>
> Now, if the NGO community really believes this is why we're here, I think we can stop this discussion right here and now. This is ridiculous … I think we should accept that the other people around this room are doing this in trust and in the right way to find solutions. I must say I feel somewhat outraged about this.
>
> (Stigson, 1998)

A few minutes later, and just after the chair had requested that speakers work together in a mood of non-adversarial cooperation, NGOs replied. A colleague of the author of the 'offending' newspaper article – and a long-time debating partner of the critical speaker from business – clarified that the article referred to a much more specific area of business opposition rather than to a general aversion to regulation ('as Mr Stigson well knows'), and challenged the assumption that the fundamental basis of the dialogues was trust:

> I think trust is not the essence of this discussion. The whole point of this discussion is how can we monitor, how can we track, how can we ensure compliance. If it were based on trust, we wouldn't even be discussing these things. I think it's a naïve and obsolescent belief, and that this industry itself has said, 'Don't trust us – track us.'
>
> (Dunion, 1998)

The dialogues resumed their collegial tone, but on the second day's session, another Major Group took a confrontational strategy. Several minutes after business association representatives had declared that voluntary self-guidelines by companies would be sufficient to achieve safe products, environmentally-friendly production and healthy working conditions, a trade union representative responded:

> Mr. Chairman, … Whether in … Bangkok, or … in Bhopal, or in the United States, we have repeatedly witnessed mass deaths from irresponsible business practices. I would like to review a particularly tragic, infuriating and frustrating case. There was an explosion exactly three years ago yesterday in a chemical plant … fifteen miles from the United Nations. It killed three managers and two workers, injured forty community residents and emergency service personnel, and contaminated the river.
>
> (Frumin, 1998)

The trade union representative then provided a devastating, step-by-step narration of the accident and a systematic analysis of the sequence of technological, management and enforcement failures that had allowed the catastrophe to take place:

> What were the problems? Problem one, the lack of stringent standards – the chemicals that exploded were not covered by … the primary US (regulation) on industry hazards, and there is no legal mandate that industry managers alert local authorities in the event of emergency. Problem two, the incompetence of corporate managers – after the plant was evacuated, the [company] vice-president ordered workers back inside. Minutes later they were dead. Problem three, improper hiring procedures – the company falsified his credentials. Problem four, failure to discipline corporate managers – he still works for the company.
>
> (Frumin, 1998)

The trade union representative itemized a total of ten technological, management and enforcement failures, including:

> government failures to investigate catastrophic incidents, and identify and close gaps in standards and enforcement, lack of national and core labor standards

and enforcement nationally and in international agreements on environment, development and trade; lack of worker participation, and lack of real oversight and effective intervention by community representatives, NGOs and local government officials.

(Frumin, 1998)

Following the lunch break, Dan Roczniak, a representative of the US-based Chemical Manufacturers Association adamantly replied, rejecting the implication that his industry as a whole was responsible for the failures of the company involved in the accident:

I would like to take a minute right now to set the record straight. [That company] has never been a participant in [our industry's safety] program – at the time of the accident, or now, or at any time.

(Roczniak, 1998)

The trade union speaker was not swayed. 'The details are unimportant. It's our unfortunate experience that the involvement that the … program has provided has been insufficient.'

Another volatile exchange over the same issue almost caused a break-up among the industry groups, when representatives of the International Chamber of Commerce (ICC) supported a call from NGOs and trade unions for a continuing multi-year review of voluntary codes of conduct after the commission meeting had concluded – an idea which the other broad business coalition at the Dialogue, the World Business Council for Sustainable Development (WBCSD), had initially opposed.

It was agreed, by the close of the CSD session, that all three civil society sectors would cooperate with governments in a multi-stakeholder review of the effectiveness of voluntary codes, facilitated by the UN Division for Sustainable Development.

Despite the contentiousness – or perhaps because of it – the result was a precedent-setting model of civil society participation. It was the first concrete output of the dialogue sessions.

Consumer guidelines

Since the Rio Summit in 1992, there had been little work done to address chapter four of Agenda 21. Norway had hosted a series of workshops just after Rio (1994), but without any real concrete outcome. One way that it was thought might help push national legislation was to amend the UN consumer guidelines for national consumer protection regulations. Compliance with these by governments is voluntary, but they are a model for consumer regulations in many nations throughout the world.

Brazil offered to host a workshop in São Paulo, and its results established a process for a future CSD session to suggest changes to national regulations on consumer protection. It requested governments to consult stakeholders nationally on adding sustainable development issues to consumer protection standards, and to report to the follow year's CSD meeting.

Freshwater

The other significant issue on the CSD agenda was freshwater. This had been a very controversial issue the previous year, and three significant intersessional meetings had been sponsored by governments in order to build on the discussion at the Special Session. Despite the efforts, no agreement was reached.

The CSD did, however, call for an assessment of the state of freshwater resources, and the debate did enable continued discussion around the issue of transboundary water sources.

Overall, the 1998 CSD produced mixed results. The year introduced the innovative dialogue sessions, but the meeting suffered because the visionary founding director of the Division of Sustainable Development, Joke Waller-Hunter, had moved on to run the Organization for Economic Co-operation and Development (OECD) and a new director for DSD had yet to be appointed.

Initiating a 2002 Summit

Frustrated by the lack of achievement a year earlier at the Rio+5 Special Session, several individuals with roots in the UK NGO community began a decidedly unofficial process to improve options for what could be the next scheduled heads of state sustainable development meeting – a Rio+10 Summit, in 2002. Derek Osborn (co-chair of the UNGASS preparatory process), Stakeholder Forum (a UK-based NGO) and the NGO CSD Steering Committee informally organized 'working dinners' for governments during the 1998 CSD meetings, to review what had gone wrong in the preparations for the Special Session and what issues might effectively be addressed at a new summit.

The organizers produced a 'non-paper' which encouraged governments to start focusing on the issue early. The UK and Germany were the first two developed countries to take a lead in considering a 2002 Earth Summit, led respectively by Sheila McCabe and Karsten Sachs. The UK provided early ministerial support for the idea. The new Labour Prime Minister, Tony Blair, had had his first international appearance at the Special Session in 1997, with a delegation that included five ministers, including the two highest officials in government.

Blair was constructing an image for himself as a 'green Prime Minister', and so was happy to lead on a possible new summit, in 2002.

Originally there was hope that India, as a developing country with a rapidly emerging economy, would host the summit in 2002. Germany and the UK had sent missions to New Delhi to discuss the possibility in late 1998. In early 1999, however, India and Pakistan became involved in a border dispute, called the Kargil War, which many worried might escalate into a nuclear exchange. This took India off the table as a possible host for the summit.

International environmental governance

Since the 1992 summit in Rio, the CSD had become the primary venue for national environmental ministers to interact. Outside the CSD, there was increasing fragmentation of the environment agenda into distinct issue areas covered by international conventions, each with its own governing board of governments and accredited stakeholders. While UNEP played a secretariat or administrative role for some of these, others were scattered among a variety of UN programmes and agencies (e.g. International Code of Conduct on the Distribution and Use of Pesticides; International Undertaking on Plant Genetic Resources, in the FAO; World Heritage Convention Concerning the Protection of the World Cultural and Natural Heritage in UNESCO). This created an increasing organizational hodgepodge, and resulted in limited cooperation among secretariats.

The appointment of Klaus Töpfer as executive director of UNEP presented an opportunity to try and sort this out. In 1998, UN Secretary-General Kofi Annan established the UN Task Force on Environment and Human Settlements to investigate structural reform to UN activities in those areas. The inclusion of human settlements – meaning primarily urban areas – came about because of conflicts the Centre for Human Settlement (CHS) was facing after the Habitat II Conference in 1996.

The Task Force would be chaired by Töpfer, who was in the unique position at that time of also being the head of the CHS, as well as Executive Director of UNEP. Known as the 'Töpfer Task Force', it included many Rio veterans, such as Maurice Strong, Gus Speth, Tommy Koh, Nitin Desai and John Ash.

The preference by supporters like Germany, South Africa and Singapore for a World Environment Organization (WEO) had failed in 1997, and it was clear in 1998 that there was still no political will to advance it. Instead, the task force made a series of suggestions to help re-establish UNEP as the primary international venue for environmental issues. The General Assembly approved several:

- establishing of an interagency group called the Environment Management Group (EMG);
- increasing UNEP's scientific and information base;
- sponsoring joint meetings of convention secretariats to ensure that their work was complementary, filled policy gaps and took advantage of potential synergy between conventions; and
- organizing an annual, ministerial-level, global environmental forum.

(United Nations, 1998)

Several of the Task Force suggestions were not approved:

- universal membership of all UN nations in UNEP; and
- engaging civil society organizations in a model similar to that in the UN CSD.

Although strengthened by the outcomes of the Task Force Report, UNEP was not given the extra political weight of a specialized agency, keeping that debate alive for Rio+10, in 2002.

Seventh Session of the Commission on Sustainable Development (1999)

In terms of generating focus and clear results, CSD 1999 may have been the most successful of all CSD sessions. The new chair was Simon Upton, Minister of the Environment of New Zealand. He immediately set up a coordinating team drawn from his own government to support the UN secretariat, under the leadership of David Taylor, his capable chief of staff. He treated the CSD work schedule like a political campaign – first analysing where there were areas of likely agreement and working to secure those. Second, isolating those areas where agreement was impossible. Finally, he worked to expand the areas of agreement into concrete outcomes and action.

This 'triage' strategy brought not only focus and results, but a surprising sense of enthusiasm and achievement.

1999 stakeholder dialogues on tourism

The issue of tourism had not been on the agenda at the Earth Summit in Rio, but it was added to the CSD's agenda at the Special Session in 1997, as evidence emerged about its impacts on pollution of the oceans and atmosphere, and exploitation of local communities and natural resources. It was scheduled as the topic of the two-day dialogues among stakeholders and governments.

Aware that tourism stakeholders had not participated at the UN previously, the chair's staff hosted a stakeholder meeting in London to review each sector's positions and anticipate the expected flow of the two-days' sessions.

The 1999 dialogues were probably the best prepared and most successful of any. They were conducted between four stakeholder groups: industry, trade unions, NGOs and local governments. Their subtopics were 'industry initiatives', 'consumer behaviour', 'sustainable development' and 'coastal impacts related to tourism.'

Both the CSD stakeholders and the chair approached the sessions determined to achieve discrete conclusions. During preparatory meetings, the NGO CSD Steering Committee presented the chair with a number of ideas for streamlining the dialogues and focusing them more on outcomes. The industry group was led by the World Travel and Tourism Council (WTTC), a broadly organized business association that had been increasingly active on sustainability and environmental issues and had developed after Rio the WTTC's *Agenda 21 for Travel and Tourism*. Trade unions (led by the ICFTU) brought a diverse delegation of hotel and transport workers and a series of well-developed policy recommendations. And officials from local authorities (led by ICLEI) brought direct understanding of the potential positives and negatives of tourism on their economies.

Upton used his prerogative as chair to unilaterally restructure the sessions to make them more interactive and dynamic. He put both stakeholders and government representatives on the spot with direct questions, requiring a response to previous speakers' statements.

The results were not only more interactive and dynamic sessions, but also more colourful – at one point the industry coalition seated a life-sized stuffed figure of a cartoon dodo in one of its speakers' chairs, presumably to personify its support for natural (or extinct) ecosystems. The sessions also yielded a concrete result – recommending that the CSD establish a multi-stakeholder ad-hoc informal open-ended working group on tourism:

> to assess financial leakages and determine how to maximize benefits for indigenous and local communities; and to prepare a joint initiative to improve information availability and capacity-building for participation, and address other matters relevant to the implementation of the international work programme on sustainable tourism development.
>
> (UNCSD, 1999)

Even more could have been achieved, but due to a disagreement within the NGO grouping an opportunity to link the 1998 and 1999 CSDs together with a review of the WTTC's Green Globe voluntary initiative was lost. Some of the tourism NGOs were worried that by simply reviewing the initiative they might bestow some legitimacy to it. Other NGOs had fought very hard the previous year to review the industry initiative both on the merits, and in order to establish precedents on civil society oversight. So for some it seemed a huge disappointment, possibly more related to personalities than to substance.

The question then was whether the recommendation would be adopted by governments. A chair's summary of the sessions incorporated the tourism working group recommendation, but there was no formal procedure for including them in the governments' negotiations. Stakeholders had been concerned that very little of what had been discussed at the 1998 dialogues was picked up into the formal debate by governments. That seemed to defeat the dialogues' purpose of helping governments take note of stakeholders' views.

Upton dramatically resolved that by entering his chair's summary of the dialogues as a set of amendments by the New Zealand government to the formal session's negotiating text. The subtle but significant effect was to ensure that governments had to address stakeholder views, and would have to actively reject those they disapproved of, rather than have to argue to add those they wanted put in. Much of the texts from the dialogues indeed survived the negotiations, most significantly the establishment of the ad-hoc open-ended working group.

Consumer guidelines

Following up on the process established by the previous year's session, the 1999 CSD heard the reports of governments' national consultations on adding sustainable development issues to consumer protection standards. The commission then negotiated an expanded set of draft consumer protection guidelines that included sustainable consumption standards. These were submitted and accepted by the General Assembly. As many countries' national laws on consumer protection are drawn from UN guidelines, this was a significant success.

Oceans and seas

In December 1998 the governments of the UK and Brazil co-sponsored the heavily attended Second London Oceans Conference. Its conclusions fed directly into the 1999 meeting of the CSD, which focused on the areas of marine pollution and over-fishing.

Its key recommendation was that the UN General Assembly establish an open-ended, informal consultative process, to broaden and deepen the negotiations of policies on oceans. This new body would be coordinated by the existing UN GA-based secretariat of the Convention on the Law of the Sea (UNCLOS). It would meet immediately prior to the annual UNCLOS session, and it would utilize CSD rules of procedure concerning civil society engagement.

This last item was a very significant development. Until then, the UN General Assembly granted stakeholders virtually no 'rights' to participate in its processes. All the environment and sustainable development forums that stakeholders had been given access to were coordinated under the Economic and Social Council, not the General Assembly. This now created an opening for stakeholders to begin to engage in a GA subsidiary forum, which had maintained a far more traditional 'governments-only' policy. It was another example of the CSD having a subtle but significant impact on the way the UN system dealt with stakeholders. The Earth Summit participatory model was continuing to expand.

The new Informal Consultative Process has now met for over a decade, and successfully initiated discussion among governments on issues including deep-sea trawling, and human-produced underwater ocean noise (often produced by military sonar and explosions used in petroleum exploration, lethally damaging fish stocks, dolphins and whales). In each, issues that at first were perceived as being marginal and drawing little support from governments grew over several years to gain majority backing thanks to their new access and steady, creative campaigning inside the process by NGOs.

Earth Summit 2002

Continuing the 'informal' government meetings on options for Earth Summit 2002, former co-chair Osborn again chaired a dinner discussion on what form the ten-year review might take, what its scope might be, what emerging issues might be addressed, and what the preparatory process might look like. NGO representatives added their support for holding a full summit, as opposed to simply a review, in their statements to the CSD plenary.

Two years of lobbying culminated with ministers approving many of the NGO suggestions, when the commission agreed to request that the Secretary-General report to the next year's CSD with suggestions on a preparatory process for a comprehensive review of the implementation of Agenda 21. The commission:

> requests the Secretary-General to present a preliminary report for discussion at CSD-8, including suggestions on the nature and scope of the preparatory process for the next comprehensive review of the implementation of Agenda 21, with a view to providing the Secretary-General with guidance for preparation of his report to the 55th session of the GA.
>
> (UNCSD, 1997)

Brocket Hall meeting in Ghana

Since 1993, an informal meeting of some environment ministers had occurred each year, held between the meetings of the CSD and UNEP Governing Council, usually in September. The initiative was started by the UK and called the Brocket Hall process, its venue alternated between a developed and developing country. In 1999 it was the turn of Ghana to host the meeting. Recognizing potential impacts on a summit, NGOs opened negotiations with Ghana's Environment Minister, Cletus Avoka, to request an exchange of views between NGOs and ministers. A session was scheduled – the first time for such a government/stakeholder meeting. With NGOs from Ghana taking the lead, the session focused mainly on Earth Summit 2002, and presented strong NGO support for the possible summit. Governments in favour were led by the UK, Germany and Sweden, and others including South Africa and Brazil were showing support. The agenda flowed from the informal process in New York and a series of 'non papers' that had been coordinated mostly by NGOs, and were now helping to mobilize government support for the event.

Meanwhile, with input from NGOs, the African group of governments considered its strategy for 2002. Realizing that the CSD's rotation process meant it would be Africa's turn to chair the commission in 2002 – and therefore also the summit – the bloc decided to select its CSD bureau member for 2002 right away, designating Cletus Avoka. At the meeting, South Africa offered itself as a possible host for the Rio+10 meeting.

By the end of the year, the CSD coordinators for local governments (ICLEI), trade unions (ICFTU) and industry (WBCSD) had endorsed the call for the summit.

Global Environmental Outlook

With the year 2000 approaching at a time of record economic prosperity, but focus on the environment and the world's poor appearing to slide, UNEP decided to frame its flagship 'state of the world' publication, the *Global Environmental Outlook* (GEO), in unusually direct and dramatic style. It launched GEO2000 in London in September 1999 with major promotion to international media. The report highlighted two over-riding trends that would impact nations in the new millennium:

> First, the global human ecosystem is threatened by grave imbalances in productivity and in the distribution of goods and services. A significant proportion of humanity still lives in dire poverty, and projected trends are for an increasing divergence between those that benefit from economic and technological development, and those that do not. This unsustainable progression of extremes of wealth and poverty threatens the stability of the whole human system and with it the global environment.
>
> Secondly, the world is undergoing accelerating change, with internationally-coordinated environmental stewardship lagging behind economic and social development. Environmental gains from new technology and policies are being overtaken by the pace and scale of population growth and economic development. The processes of globalization that are so strongly influencing social evolution need to be directed towards resolving rather than aggravating the serious imbalances that divide the world today. All the partners involved – governments, intergovernmental organizations, the private sector, the scientific community, NGOs and other major groups – need to work together to resolve this complex and interacting set of economic, social and environmental challenges in the interests of a more sustainable future for the planet and human society.
>
> (UNEP, 1999)

The report gained significant print and television coverage, and helped reinvigorate momentum on climate and sustainable development going into the new millennium.

The launch also featured a session on Earth Summit 2002 and what it should address – the first time the media were really hearing about the possibility of a new summit.

Eighth Session of the Commission on Sustainable Development (2000)

The new millennium opened with the world largely at peace, with what most indicators registered as record economic growth, and with major economies thriving. But dissent was bubbling to the surface over increasing disparities in wealth and the increasing predominance of multinational corporations and international trade. The previous November, a World Trade Organization (WTO) meeting meant to approve an expanded regimen of global trade had degenerated into the 'Battle of Seattle' – four days of at-times violent confrontations between police and demonstrators. Those activists represented a new coalition of developing country advocates, international environmentalists and developed-country trade unions.

Their criticisms were many – lack of transparency, lack of involvement by developing countries in decision making, lack of access by stakeholders. But for developing countries, the key issue was the blatant imbalance of the trade agreements:

> developing countries have remained steadfast in their demand that developed countries honor Uruguay Round commitments before moving forward full force with new trade negotiations. Specifically, developing countries are concerned over developed countries' compliance with agreements on market access for textiles, their use of anti-dumping measures against developing countries' exports, and over-implementation of the WTO Agreement on Trade Related Aspects of Intellectual Property Rights (TRIPs).
>
> (ICTSD, 1999)

Responding to the potential strength of this emerging opposition, US President Clinton had two weeks earlier issued an executive order which committed the United States to a policy of 'assessment and consideration of the environmental impacts of trade agreements,' and stated that '[t]rade agreements should contribute to the broader goal of sustainable development' (Clinton, 1999).

Whether the world's sole remaining superpower and other governments would act on such principles in negotiations would remain to be seen.

The intergovernmental community responded, as the UN Secretary-General led governments in planning a Millennium Development Summit to focus resources on achieving specific poverty reduction targets. So as the CSD met to deal with an agenda that included trade, finance and investment, the timing couldn't possibly have been more perfect – or more difficult.

The new chair of the CSD would be Juan Antonio Mayer, Colombia's dynamic Minister of the Environment who seemed to represent an active new Latin American diplomatic generation. He had overseen the successful negotiations for the Cartagena Bio-safety Protocol under the Convention on Biological Diversity (CBD) in 1999. He took a similar approach to that of Simon Upton and brought in a skilled team to support the preparation of the negotiations.

Multi-stakeholder dialogues on agriculture

The 2000 CSD dialogues provided the first opportunity for participation by the major group representing farmers. All four themes revolved around sustainable agriculture. One also focused on globalization and trade:

- choices in agricultural production techniques, consumption patterns and safety regulations –with regard to the potential opportunities and threats to sustainable agriculture;
- best practices in land resource management to achieve sustainable food cycles;
- knowledge for a sustainable food system – identifying and providing for education, training, knowledge sharing and information needs; and
- globalization, trade liberalization and investment patterns – economic incentives and framework conditions to promote sustainable agriculture.

For the third successive year, the dialogues produced a concrete outcome involving civil society. FAO and the CSD secretariat were invited to set up an ongoing multi-stakeholder dialogue on Sustainable Agriculture and Rural Development (SARD). It would emphasize 'best practice' cases studies, and prepare the issue for the Earth Summit in 2002.

Unlike previous dialogues, the chair did not look for areas of consensus, but instead looked for areas of critical difference – ways to highlight the work that was still needed. The model had supporters as well as detractors. It might have worked better had the mood been more constructive among governments.

Agriculture

The CSD took an important step in establishing SARD as the framework for approaching agriculture in the new century. The Commission reaffirmed the commitments made at the World Food Summit in 1996 for the goal of reducing by half the number of undernourished people by 2015. This would become one of the Millennium Development Goals (MDGs) later in 2000. There was support for governments completing negotiations on the International Undertaking on Plant Genetic Resources for Food and Agriculture, and also to ratify the recently negotiated Cartagena Bio-safety Protocol.

Forests

One of the successes of the 2000 CSD was the recommendation to set up a UN Forum on Forests. This brought to a conclusion the process initiated in 1994 by establishing a subsidiary body of ECOSOC with the primary objective to promote: 'the management, conservation and sustainable development of all types of forests and to strengthen long-term political commitment'.

Its work would be based on the Rio Declaration, the Forest Principles, chapter 11 of Agenda 21 and the outcome of the IPF/IFF Processes, and other key milestones of international forest policy.

It was another example of the CSD taking an issue as far as it could and then establishing a permanent forum to take the next steps – very similar to what it had done for oceans and seas.

Financial resources and mechanisms

Despite a decade of global economic boom, and obvious concerns about globalization, governments at CSD 2000 were unable to reach any meaningful agreement on finance. The developed countries were reminded of their commitments made in Rio, and although ODA had stabilized at a slightly higher US$53 billion–US$54 billion, it was still well under the US$60 billion in 1992 and nowhere near the US$120 billion Agenda 21 said was needed for its implementation. The US was taking a particularly hard line, saying that aid flows were unlikely to go up again, and that this would have an implication in implementing Agenda 21.

This was a long way from Buff Bohlan's promise as the US negotiator in 1992 to help mobilize new resources, and deeply ironic considering that there was a Democratic president now in the White House. The EU took a much softer line which would play significantly into the MDG summit later that year, citing the need to unify around the 'eradication of poverty through sustainable development in the framework of the international development targets derived from UN Conferences and Summits'.

There was no agreement in key areas such as trade-distorting and environmentally harmful subsidies, trade liberalization and the Clean Development Mechanism (CDM) under the UNFCCC. In a typical diplomatic finesse that resolved nothing, globalization was simply recognized as a set of opportunities, risks and challenges.

UNEP and governance – The Malmö Declaration

Following the Töpfer Task Force, the first meeting of the new UNEP Global Ministerial Environmental Forum (GMEF) was held in Malmö, Sweden in May 2000. The resulting Malmö Declaration provided a mandate for UNEP's role in preparation for the Earth Summit in 2002, and for continuing the discussion on UNEP reform that had been started by the Task Force. It wouldn't be until 2010 that the UNEP GMEF would again negotiate a declaration, in that case for the summit in 2012.

Earth Summit 2002

In March 2000, over 70 representatives of governments, intergovernmental agencies and Major Groups met in Sussex, England, for the 598th Wilton Park Conference to discuss preparations for the summit in 2002. Hosted by Wilton Park and Stakeholder Forum, the event included Cletus Avoka, the Environment Minister of Ghana, who most assumed would chair the summit process, and Nitin Desai, Under Secretary-General for UN DESA and soon to be Secretary-General of the 2002 summit.

The event approved a 'non paper' and recommended that:

- the summit be called the World Summit on Sustainable Development (WSSD);
- the review of implementation of Agenda 21 should be separated from the forward looking summit itself;

- the summit should not be held in New York, but in a developing country;
- multi-stakeholder processes should be integrated into all international preparations for the summit;
- the summit should provide a focus on poverty eradication;
- the summit should advance equitable sustainable consumption and production practices; and
- the summit should address the impact of globalization on sustainable development and should establish a new financial compact (to become known as the new deal) between developed and developing countries including agreement on how to resource the implementation of Agenda 21, the transfer of appropriate technology and agreed timetables.

(Dodds, 2000)

At the 2000 CSD not one, but four countries – South Korea, Brazil, Indonesia and South Africa – offered to host the summit, the first time that had ever happened in a UN process. South Africa had been actively preparing to offer to host the summit, and brought a large team to the preparatory meeting. This caused real problems for the CSD secretariat. Until a General Assembly resolution declared there would be such a conference, and a host country was agreed to, the secretariat was not able to be active in any summit preparation.

It took until December before an agreement could be reached, costing valuable time. Surprisingly, the compromise was made possible by national elections in Ghana. The Ghanaian Environment Minister had been actively preparing for being chair of the preparatory process. In the December 2000 Ghanaian vote, however, opposition parties won control of government via the ballot box for the first time in African history. The defeat of the governing party cost the minister his post and the potential chair of the summit. But it paradoxically enabled a deal to emerge between South Africa and Indonesia in which South Africa was selected as the summit host country, and Indonesia was selected to hold the final WSSD preparatory meeting and to serve as its chair.

Millennium Development Summit

Even as government and NGO sustainable development advocates struggled to gain momentum for a summit, diplomatic attention was rapidly building for an entirely new process. In September 2000, the largest number of heads of state ever to attend any meeting assembled in New York for the Millennium Development Summit. Empowered – or embarrassed – by five years of soaring growth among the economically wealthy nations, they agreed a set of eight goals with twenty-one very specific targets.

To many advocates, the overtly limited targets of the MDGs were an example of the failure of governments to deliver on the much broader conference commitments of the 1990s. But supporters hoped that by simplifying the commitments, developed

Box 6.2 **Millennium Development Goals**

- Goal 1: Eradicate extreme poverty and hunger
  - Target 1A: Halve the proportion of people living on less than US$1 a day
  - Target 1B: Achieve decent employment for women, men and young people
  - Target 1C: Halve the proportion of people who suffer from hunger
- Goal 2: Achieve universal primary education
  - Target 2A: By 2015, all children can complete a full course of primary schooling, girls and boys
- Goal 3: Promote gender equality and empower women
  - Target 3A: Eliminate gender disparity in primary and secondary education preferably by 2005, and at all levels by 2015
- Goal 4: Reduce child mortality rate
  - Target 4A: Reduce by two-thirds, between 1990 and 2015, the under-five mortality
- Goal 5: Improve maternal health
  - Target 5A: Reduce by three quarters, between 1990 and 2015, the maternal mortality ratio
- Goal 6: Combat HIV/AIDS, malaria and other diseases
  - Target 6A: Have halted by 2015 and begun to reverse the spread of HIV/AIDS
  - Target 6B: Achieve, by 2010, universal access to treatment for HIV/AIDS for all those who need it
  - Target 6C: Have halted by 2015 and begun to reverse the incidence of malaria and other major diseases
- Goal 7: Ensure environmental sustainability
  - Target 7A: Integrate the principles of sustainable development into country policies and programmes; reverse loss of environmental resources
  - Target 7B: Reduce biodiversity loss, achieving, by 2010, a significant reduction in the rate of loss
  - Target 7C: Halve, by 2015, the proportion of the population without sustainable access to safe drinking water
  - Target 7D: By 2020, to have achieved a significant improvement in the lives of at least 100 million slum-dwellers
- Goal 8: Develop a global partnership for development
  - Target 8A: Develop further an open, rule-based, predictable, non-discriminatory trading and financial system
  - Target 8B: Address the Special Needs of the least-developed countries (LDC)
  - Target 8C: Address the special needs of landlocked developing countries and small island developing States
  - Target 8D: Deal comprehensively with the debt problems of developing countries through national and international measures in order to make debt sustainable in the long term
  - Target 8E: In cooperation with pharmaceutical companies, provide access to affordable, essential drugs in developing countries
  - Target 8F: In cooperation with the private sector, make available the benefits of new technologies, especially information and communications

countries would at last provide the resources and the political support to see them delivered.

Most environmental NGOs had decided to pass up the Millennium Development Summit, having made an early assessment that it would not be an important meeting. As a result, the sustainable development section of the goals and targets were weakly constructed. At first, in the summer of 2000, as negotiations became bogged down, it looked like the pessimists might be right. In the end, however, the Millennium Development Summit was considered perhaps the most significant UN event in its history, because for the first time aid flows would dramatically increase and the possibility of goals being achieved became more of a reality.

The initial lesson was that very specific and clearly understandable agreements may have greater hope for implementation than broader, more comprehensive agreements – even if those would promise greater results in principle. Ten years later, those results are less clear. While making significant progress on some goals such as MDG1 (poverty) and MDG4 (child mortality), the MDGs have run into all-too-familiar failures to follow through in areas such as MDG7 (water) and MDG3 (gender equality).

Earth Summit 2002 – election set-backs

A critical element in the optimistic 2002 scenario cultivated by summit advocates was the 2000 US presidential election between Vice-President Al Gore and Governor George W. Bush. Gore had seemed a shoo-in as he and President Clinton presided over a robust economy that was running a surplus, at a time of peace and tranquility. Many in the sustainable development community were looking forward to a President Al Gore attending the WSSD and confidently addressing issues like climate change, trade and alleviation of poverty. After all, he had attended the Earth Summit in 1992 as the author of the widely-respected *Earth in the Balance* and leader of the US senatorial delegation, and hopes were high that he'd arrive in 2002 in Johannesburg positioned to offer global leadership. But it wasn't to be. The stunning and chaotic American election changed the entire landscape of what might be possible at the summit. Until then, the US had been supportive of a discussion on a new deal between developed and developing countries. That plan now lay in ruins in the recounts of Florida.

In Europe, a similar election defeat removed from office the environment and energy Minister of Denmark, Svend Auken, a dynamic and politically effective advocate of both environmental and economic issues. With his country scheduled to assume the rotating presidency of the EU in July 2002, Auken was also due to play a leading role in WSSD negotiations.

So instead of appearing like a venue of potential political triumph, the summit suddenly risked looking like a diplomatic disaster zone.

Ninth Session of the Commission on Sustainable Development

The election to CSD chair of Bedřich Moldan of the Czech Republic was welcomed by many. He had played a significant role in the Earth Summit in 1992 and his election continued the pattern of the chair being held by current or former ministers – one indicator of high level national support for the commission. By now, however, the commission was becoming less and less effective, and needed drastic reorganization.

To add to its problems, the whole CSD was overshadowed by planning for the WSSD, whose first preparatory session would meet immediately after the commission's session in May.

Stakeholder dialogue on energy and transport

In the dialogue's fourth year, two of the three issues added to the sustainable development agenda at the Special Session were to get an airing. The dialogue was between NGOs, business and industry, workers and trade unions, local authorities, and the scientific and technological community. The four sessions of the dialogues addressed:

- achieving equitable access to sustainable energy;
- sustainable choices for producing, distributing and consuming energy;
- public–private partnerships to achieve sustainable energy for transport; and
- sustainable transport planning: choices and models for human settlement designs and vehicle alternatives.

Unlike the previous three dialogue sessions, there was no process set up to take forward any of the outcomes.

Energy

The Ad-hoc Open-Ended Intergovernmental Group of Experts on Energy and Sustainable Development – the two-year process set up to prepare the discussion – made recommendations, but it was clear that any real progress on energy would have to wait until the WSSD. There was no clear direction or additional resources agreed to help the transition of developing countries towards more sustainable energy sources. There was even resistance to the application of Principle 16 of the Rio Declaration (internalization of environmental costs) in establishing energy policies.

Financial resources and mechanisms

ODA had slipped again, this time to below US$53 billion, and developing countries reminded developed countries not only of their commitments to sustainable development but to the Millennium Development Goals agreed the previous

September. The reality was that those MDG commitments would take a couple of years to work their way through governments' budgeting systems. The impacts of the MDG Summit on ODA would be positively felt over the next decade, but the mood one year before the 2002 summit was increasingly gloomy.

UNEP Governing Council

At the UNEP Governing Council in February 2001 the outcomes from the Malmö Declaration were incorporated and UNEP created an open-ended intergovernmental group of ministers to advance the discussion on strengthening the structure of international environmental governance (IEG). It would then feed into the WSSD.

9/11

One year before the scheduled summit in Johannesburg, the intergovernmental world was stunned by the terrorist attacks of 11 September 2001, and all assumptions of the diplomatic community were shattered.

To say that 9/11 changed everything is an understatement. After an initial outpouring of support for the victimized United States, what initially looked like a potentially sincere coming together of the world to address the root causes of conflict, instead became 'are-you-with-us-or-against-us?' polarization. A summit being organized against this backdrop saw its horizons shrinking very quickly. Money that had been impossible to obtain for development aid suddenly became easy to obtain for taking military revenge. A plunging stock market undermined any effort to organize partnership agreements with previously interested corporations. Many in NGOs and in governments argued that by more directly dealing with underlying poverty and growing inequality, and the resulting feelings of hopelessness and alienation, governments and the summit could make a critical impact and address issues at the core of both terrorism and sustainable development.

The Heinrich Böll Foundation generated a debate and a publication, 'The Road to Johannesburg after September 11, 2001,' which tried to open a more profound discussion about 9/11 and sustainable development.

> If September 11 has indeed helped people understand that terror arises from poverty and injustice in the Third World, and that terror can strike any of us anywhere, then perhaps it is possible to turn the tragic events to good purpose on our road to Johannesburg.
>
> (Brusasco-Mckenzie, 2002)

But conference organizers seemed immobilized – unable to act on even the most obvious questions. It took until April 2002 for the WSSD secretariat to decide to reschedule the summit so that its closing plenary session – attended by dozens of heads of state and government – would not coincide with the precise first anniversary of the 9/11 attack.

Chapter 7

Johannesburg – dealing with new realities

Delegates poured into the city's Sandton Conference Centre in extraordinary number. The Johannesburg Summit was similar in size and complexity to the Rio 1992 Earth Summit, with more than 10,000 government delegates, 104 heads of state or government, 8,000 accredited NGOs and 4,000 media representatives. A multitude of exhibitions, meetings and side-events ran non-stop around Johannesburg and other South African cities drawing thousands of additional local participants, non-accredited NGOs and press. The UN estimated that over 40,000 people took part in the summit in one way or another.

The conference had been organized with the sense of excitement and positive anticipation that always accompanies an intergovernmental summit, even if its prospects for success aren't clear. But during the two years of preparatory meetings, a series of deeply significant external events had profoundly shifted the political landscape around the conference, and instead of flying into a horizon of cooperative opportunity, participants arrived to face an intense fractious battlefield.

The intensity and frenetic activity were there, but the atmosphere seemed significantly sombre and subdued.

The preparatory process

The process leading up to the WSSD had started in February 1998 with informal dinners, workshops and quiet conversations. Governments, NGOs, UN officials, former CSD chairs and UN officials all agreed there was a need for a summit to 're-boot' sustainable development. It was clear to them that there would need to be a serious attempt to address the broken promises of the previous decade, particularly funding for sustainable development implementation in developing countries.

As the institutional process kicked in, the focus of the summit was agreed by the UN General Assembly as being to:

- undertake the comprehensive review and assessment of the implementation of Agenda 21 and the other results of the conference on the basis of the results of national assessments and sub-regional and regional preparatory meetings, the documentation to be prepared by the Secretary-General in collaboration

Box 7.1 **WSSD preparatory process**

National preparatory process 2000–2001

Governments used a number of avenues to address the call for national progression reports. Some utilized national councils on sustainable development (NCSD) where they existed or created bodies for the WSSD process.

Regional preparatory process

The UN Regional Economic Commissions were tasked with creating regional assessments. The regional meetings in Europe (24–25 September 2001), Latin America (23–24 October 2001), Africa (15–18 October 2011), West Asia (24 October 2001) and Asia-Pacific (27–29 November 2001) engaged stakeholders in a dialogue around the assessments.

Global preparatory process

PrepCom I (New York 31 April to 2 May 2001)

Organizational matters – focused on the election of the bureau and the summit's organization of work.

PrepCom II (New York 28 January to 8 February 2002)

This preparatory session identified key issues for the summit.

PrepCom III (New York 25 March to 5 April 2002)

A chairman's draft was produced – released originally in February for governments and stakeholders to prepare. South Africa floated a 'non-paper' to reorganize the conversation. The chair, Emil Salim, produced a compilation document, but not in the structure suggested by South Africa.

PrepCom IV (Bali 27 May to 5 June 2002)

The chairman's text, released on May 13, did not address what governments had hoped it would. The UN Secretary-General moved into high gear and launched a focus approach around water, energy, health, agriculture and biodiversity (WEHAB). PrepCom VI developed guidelines for Type II initiatives. The WEHAB agenda turned out to be too late to refocus the summit, therefore prompting governments to move through the chairman's text.

continued...

Box 7.1 continued

World Summit on Sustainable Development
(Johannesburg 26 August to 4 September 2002)

There were three main outcomes from WSSD:

- political declaration: this was negotiated behind closed doors with no stakeholder involvement (Type 1: negotiated documents with principles, commitments and targets by governments);
- plan of implementation: this was the negotiated chair's text and adopted on the last day of the conference (Type 1);
- partnerships and initiatives: this included 'coalitions of the willing' – involving two or more stakeholder groups and governments and civil society partnerships towards implementation of sustainable development (Type 2: commitments, targets and partnerships by individual or groups of governments and stakeholders).

---

with the task managers, and other inputs from relevant international organizations, as well as on the basis of contributions from major groups;
- identify major accomplishments and lessons learned in the implementation of Agenda 21;
- identify major constraints hindering the implementation of Agenda 21, propose specific time-bound measures to be taken and institutional and financial requirements, and identify the sources of such support; and
- address new challenges and opportunities that have emerged since the conference, within the framework of Agenda 21.

(United Nations, 2000)

The preparatory process was scheduled for 40 formal negotiating days, over a period of nearly two years. Governments were therefore back to a model closer to the Earth Summit, with significant opportunity to debate and resolve strategies for action.

What's the structure?

One of the main organizational issues for the summit process was the issue of the structure of the outcome document. Very early in PrepCom II, NGOs raised the suggestion that a working group should be set up to discuss the structure.

Clearly Agenda 21 had a very clear structure as a document:

- Basic for Action;
- Objectives;
- Activities;

- Means for Implementation; and
- Costings.

Some have argued that it would be best to return to this structure, as it offers a coherent approach to the issues under review. The chapters of Agenda 21 had been gathered into the following overarching sections:

- Section 1 – Social and Economic Dimensions
- Section 2 – Conservation and Management of Resources for Development
- Section 3 – Strengthening the Role of Major Groups
- Section 4 – Means of Implementation.

The problem with the Johannesburg Plan of Implementation (JPoI) was that the initial text from the chair was disjointed and had no clear narrative or proper sections. In addition, some countries didn't want the JPol to even consider sections on issues such as energy, an obstructionism effectively led by OPEC.

Some countries were convinced that a consensus on structure would emerge in PrepCom III, and suggested that NGOs were being alarmist. No such structure did emerge, except within the South African 'non paper'. It suggested that issues be addressed under a framework consisting of:

- proposed targets and timeframes;
- proposed actions;
- resources;
- institutional mechanisms;
- coordination;
- monitoring;
- stakeholder involvement; and
- implementation plan sustainability.

Governments left PrepCom III mostly in agreement with the South African suggestion. Two weeks later, however, WSSD chair, Emile Salim, circulated a draft with a more traditional approach. His structure consisted of:

(i) Introduction – including: The United Nations Conference on Environment and Development (UNCED); sustainable development for all; partnerships; good governance; peace, security, stability and respect for human rights and fundamental freedoms;

(ii) Poverty eradication – including: eradicating poverty; drinking water and sanitation; energy services for sustainable development; contribution of industrial development to poverty eradication and sustainable natural resource management; slum dwellers and child labour;

(iii) Changing unsustainable patterns of consumption and production – including: sustainable consumption and production; corporate environmental and social responsibility and accountability; sustainable development in decision-making; energy; transport; minimizing waste and maximizing reuse, recycling and use of environmentally friendly alternatives; sound management of chemicals;

(iv) Protecting and managing the natural resource base of economic and social development – including: water; conservation of oceans, seas, islands and coastal areas; vulnerability, risk assessment and disaster management; climate change; agriculture; United Nations Convention to Combat Desertification; mountain ecosystems; sustainable tourism development; biodiversity; forests; mining, minerals and metals;

(v) Sustainable development in a globalizing world;

(vi) Health and sustainable development;

(vii) Sustainable development of Small Island Developing States;

(viii) Sustainable development for Africa;
Other regional initiatives – including: sustainable development in Latin America and the Caribbean, Asia and the Pacific, West Asia, and Economic Commission for Europe regions;

(ix) Means of implementation;

(x) Institutional framework for sustainable development;

(xi) Strengthening the institutional framework for sustainable development at the international, regional and national levels, as well as participation of major groups.

With the relevant paragraphs being collected under these headings, it made for a very disjointed and incoherent document. Further, not once did it actually address the name it was given: The Johannesburg Plan of Implementation. There was no plan and minimal focus on implementation.

A new deal

During these early informal meetings, the idea of a 'new deal' on finance was first suggested. The non paper of 2000 said:

> a new deal on finance – enabling a deal on sustainable development.
> (Stakeholder Forum, 2000)

A new 'Global Deal' between governments would agree to provide the financing to implement sustainable development promised in the 1992 agreements at Rio. It would include the outcomes of the Doha Development Agenda, the Monterrey Consensus and the World Food Summit. And it could help fund achievement of the MDGs.

Derek Osborn, former co-chair of the preparatory process for Rio+5, in 1997 underlined a number of key objectives for WSSD. These 'Osborn objectives' included:

- A revitalized and integrated United Nations system for sustainable development;
- A new deal on finance – enabling a deal on sustainable development;
- an integration of trade and sustainable development;
- a clearer understanding of how governments should move forward nationally on implementing Agenda 21;
- a new charter which could lay the foundations for countries to frame their sustainable development; and
- a clear set of commitments to implement agreed actions by the UN governments and major groups.

(Osborn, 2000)

In 2001, the phrase 'Global New Deal' was picked up by Denmark, who would hold the European presidency during WSSD. The host country, South Africa, also decided to push for the idea. In June 2001, just after the first PrepCom, the Council of the European Union announced that they would seek a 'global deal' at WSSD.

Thus there was the beginning of a strategy that might be able to be build sufficient political momentum to overcome opposition to securing the funding for any outcomes from WSSD.

Such opposition was likely, but it had not yet coalesced. At the time of the first preparatory meeting in May 2001, it was still unclear which way the US would go. The new political appointments of President Bush were still going through the approval process, and while those for environment and sustainable development were not known as being particularly hostile to sustainable development cooperation, they were far from the most powerful voices in the incoming administration.

Preparatory Committee I

The first preparatory meeting dealt with the modalities for the summit.

In opening the preparatory meeting Nitin Desai, Secretary-General of the summit, presented three challenges:

Have we really come to grips with the implementation of sustainable development, do we really know what it means in operational terms?

Changes have taken place, which we cannot but take into account when we meet again at the Johannesburg Summit. The most important of these is globalization.

Finally, there is the importance of connecting the sustainable development agenda with the emerging agenda on poverty eradication. Much of the agenda on poverty eradication is people centred. It focuses attention on services to be delivered to individuals. What sustainable development can contribute to this agenda is a focus on the resources dimension.

(Desai, 2001)

As with the Rio summit, governments were asked to mobilize stakeholder views at the national level, which would feed into regional preparatory meetings in the autumn of 2001 with the first substantial PrepCom in January 2002. The summer was spent by many stakeholders engaging in their national consultations on the broad themes of the summit.

The foundation shifts

The European regional preparatory meeting on 24–25 September was just thirteen days after 9/11.

The opening statement by Klaus Töpfer, Executive Director of UNEP, revealed how completely the conversation had shifted.

> Less than two weeks ago the world was shocked by a terrorist attack which took the life of almost 3000 innocent people, and threatened the very foundations of modern society and democracy.
>
> Fighting and preventing terrorism now requires our common attention and commitment. Its challenges are new and manifold.
>
> On our way to the World Summit on Sustainable Development in Johannesburg next year we should not shy away from the consequences, but explore ways to see how we can support the unfolding worldwide campaign against terrorism. There can be no justifications at all for terrorism.
>
> (Töpfer, 2001)

Just as importantly, he concluded:

> We know that the combination of poverty and environmental degradation contributes significantly to feelings of marginalization and despair. Therefore, tackling these underlying causes will help to promote sustainable development and responsible prosperity for all people around the world. It will make the world a safer place for all of us and for future generations to come.
>
> (Töpfer, 2001)

The outcome from the European PrepCom was agreement that the summit should address these central issues:

- poverty eradication;
- sustainable management and conservation of the natural resource base;
- making globalization work for sustainable development;
- improving governance and democratic processes at all levels;
- financing sustainable development;
- education, science and technology for decision-making;
- environment and health; and
- sustainable consumption and production patterns.

It also agreed that special attention should be given to:

- possible launching of a process for a 'Global Deal';
- volume of ODA; and
- precautionary principle and living modified organisms.

In response, the Northern Alliance for Sustainability (ANPED) on behalf of 80 NGOs at the PrepCom said:

> NGOs therefore welcome the proposal for a Global Deal and look forward to working with states to further define this initiative.
>
> (ANPED, 2001)

Even with the aftermath of 9/11, the European meeting had managed to provide strong direction for WSSD.

As the initial wave of international sympathy and support poured in to the United States, there were hopes that such a disaster would bring countries closer. After all, 54 other countries had lost nationals that day. It even occurred to some that a world summit already scheduled to meet virtually one year after the attack could provide an extraordinary platform for national leaders to express – and act on – a determination to reduce poverty and marginalization, and increase security for all people.

But within weeks the US began moving in the opposite direction, toward increased polarization and confrontation.

The dashed hopes were perhaps best expressed by actor Tim Robbins in Washington, in 2003:

> I imagined our leaders seizing upon this moment of unity in America, this moment when no one wanted to talk about Democrat versus Republican, white versus black, or any of the other ridiculous divisions that dominate our public discourse. ...
>
> And then came the speech: You are either with us or against us. And the bombing began. And the old paradigm was restored as our leader encouraged us to show our patriotism by shopping and by volunteering to join groups that would turn in their neighbor for any suspicious behavior.
>
> (Robbins, 2003)

And in WSSD negotiations on the 'New Global Deal' the Bush administration revealed its attitude toward cooperation. The new administration would have nothing to do with it.

The African Regional Meeting a month later already started to factor in the aftermath of this:

> We call on the Summit to agree on what we may call the 'Johannesburg Vision': a practical expression of the political commitments made by the

international community in the Rio principles and Agenda 21, and the Millennium Declaration. These commitments envisage a global consensus on the eradication of poverty and global inequality. The World Summit on Sustainable Development provides a unique platform for the realization of this vision and must adopt a results-orientated, Johannesburg Programme of Action with clear time frames and specific targets. For the effective achievement of this programme, concrete global partnerships between governments on the one hand, and between governments, business and civil society on the other hand, are required. We believe that, through these outcomes, the Summit will provide practical meaning for the achievement of the hopes of the African Century.

(African ministerial statement, 2001)

The Latin American and Caribbean regional meeting held towards the end of October was a little stronger when it called for the need to:

address the need to explore innovative and more effective mechanisms for financing the protection of national public goods that afford global benefits and that they propose means of linking the environmental dimension with countries' fiscal policies in order to effectively incorporate financial sectors into the effort to achieve sustainable development goals.

(United Nations, 2002a)

The Asian regional preparatory meeting occurring at the end of November 2001 saw a return to more traditional language for financing sustainable development. The Phnom Penh Platform for Sustainable Development for Asia and the Pacific didn't mention the 'Global Deal':

We welcome practical and innovative ways of mobilizing and identifying resources to support developing country efforts to achieve sustainable development. In order to move forward, we therefore urge all the developed countries to strive to reach the accepted UN ODA target of 0.7% of their GNP as soon as possible. In addition, developed countries are called upon to pursue the transfer of environmentally sound technologies and know-how to developing countries on favorable terms in accordance with Agenda 21.

(United Nations, 2001)

Although still pursued by NGOs and some governments, the possibility of a major agreement between developed and developing countries that the 'New Deal' could have brought about was effectively dead. It was paradoxical that something that might have had a profound change for the better to address the root causes of poverty and some might say terrorism was destroyed by that act of terrorism on that day in September.

It was a bitter example of extremism – on both sides of the political spectrum – trumping attempts at constructive moderation.

Box 7.2 **WSSD Pledges**

- Asian Development Bank
  - US$5 million to UN Habitat and US$500 million in fast-track credit for the Water for Asian Cities Programme.
- European Union
  - US$700 million Partnership Initiative on Energy
  - US$80 million committed to the replenishment of the GEF
- United States
  - US$970 million over the following three years in sanitation and water projects
  - US$43 million to be invested in energy in 2003
  - US$2.3 billion through 2003 on health
  - US$90 million in 2003 for sustainable agriculture programmes
  - US$53 million for forests between 2002–2005

The financing that was finally announced at Johannesburg was far less than what all agreed was needed. The EU did make a timetable for how its member states would aim to reach the 0.7 per cent of GDP level by 2010. This target has not yet been met, and underlines the continued distrust by developing countries.

Governance

If the Stockholm Conference in 1972 gave the world UNEP, and the Rio Conference in 1992 created the UN Commission on Sustainable Development, the question was what the Johannesburg Summit would do for sustainable development governance.

Since Klaus Töpfer had taken over as Executive Director of UNEP, he had been strengthening UNEP organizational capacity and political space. UNEP saw Johannesburg as an opportunity to accelerate that.

Legal and regulatory frameworks

Very early on, NGOs saw the opportunity to utilize the summit to persuade governments to ratify a series of agreements that were framed as the 'Rio Conventions'. These included:

- The UN Agreement on Straddling and Highly Migratory Fish Stocks
- The Rotterdam Convention on Prior Informed Consent (PIC)
- The Procedures on Persistent Organic Pollutants (POPs)
- The UN Convention to Combat Desertification (UNCCD)
- The Kyoto Protocol (UNFCCC)
- The Cartagena Protocol of the Convention on Biological Diversity (UNCBD).

NGOs still seemed to have some clout – if only to shame governments. At the fourth PrepCom, the stakeholder newsletter *Outreach* printed 'mock soccer league tables', grading governments' performance ranging from premier league to non-league. It awarded 3 points for ratification, and 1 for signing these key conventions. The results clearly illustrated the state of convention follow-up by governments. One internally amusing result was to see that Indonesia, who was hosting the final preparatory meeting for the WSSD, was higher than South Africa. By the time delegates arrived at Johannesburg two months later, that was no longer the case.

There was also a push during WSSD for the CBD to develop a new protocol to 'promote and safeguard the fair and equitable sharing of benefits out of the utilization of genetic resources'. The Nagoya Protocol On Access To Genetic Resources And The Fair And Equitable Sharing Benefits Arising From Their Utilization To The Convention On Biological Diversity – took another eight years to negotiate.

There had been hope that WSSD would provide a venue to encourage progress towards the ratification of the Kyoto Protocol. By the start of the summit, commitments to ratify were made by Canada, Mexico and, significantly, the wavering but critically necessary Russia. It had taken Russia two years more than they had said, but without WSSD it probably wouldn't have happened.

One of the disappointments of WSSD was the lack of any movement towards clustering the Multilateral Environmental Agreements (MEAs). The UNEP-sponsored 'Cartagena Process on International Environmental Governance' papers had suggested clustering around a number of themes such as:

- biodiversity/species;
- oceans and seas;
- chemicals and hazardous wastes;
- nuclear energy and testing of nuclear weapons;
- energy/climate change/air; and
- freshwater and land related conventions.

(Dodds, 2001)

This lack of progress would be picked up as a critical issue in the following ten years, starting with chemicals and biodiversity.

Institutional strengthening and capacity building

One of the most critical issues for developing countries, in addition to funding, was capacity building for sustainable development. The summit in Rio had launched a very effective programme under UNDP called Capacity 21. One of the criticisms was that there had been a lack of linkages between the sustainable development strategies and the poverty reduction strategies at the national level. To address this, UNDP launched a revamped version of Capacity 21, which they called Capacity 2015. This would enable those activities to link into the post MDGs. Within five years, however, the programme had died due to disagreement within UNDP.

Trade and finance and development

There have been three major trade rounds since the 1972 UN Conference on Human Environment.

- Tokyo Round (1973–1979);
- Uruguay Round (1986–1994); and
- Doha Round (started 2001).

A perennial obstacle blocking efforts to resolve the critical issues of trade and financing at WSSD was the primacy of the WTO. This reflected the ongoing challenge to sustainable development negotiations presented by their older 'big brother' in the UN systemic family of negotiations.

Since the start of the environment and development negotiating process, the sustainable development agenda had, in many ways, been in competition with international trade negotiations. They had also been occurring in parallel with each other in an almost eerie way.

Indeed it was governments' sudden stampede of interest to completing the Uruguay Round of negotiations on international trade (GATT), concluded in 1994 with the establishment of the World Trade Organization (WTO), that rapidly eclipsed momentum towards implementing the 1992 Earth Summit's principles and ended what participants had thought would be a transition to an era of environment-based policy and action – the 'new organizing principle' written about in 1996 by Al Gore.

The Doha Development Round or Doha Development Agenda (DDA) was started as the trade-negotiation round of the WTO in November 2001. It focused on agriculture, industrial tariffs and non-tariff barriers, services, and trade remedies. It perhaps represented the first time developing countries were really organized for what had predominately been a developed country's country club. There was disagreement across the whole set of issues but perhaps most clearly over agriculture. The agreement had a date of 2005 to finish this round. This turned out to be impossible; as we go to press in 2012 the Doha development round is still not resolved.

The other important conference in 2002 was the United Nations International Conference on Financing for Development Conference in Monterrey. The outcome was what is known as the 'Monterrey Consensus'. Over 50 heads of state and 200 ministers of finance, foreign affairs, development and trade participated in the event with the IMF, the World Bank and the World Trade Organization (WTO) as well as key stakeholders. The outcome was agreements on debt relief, fighting corruption, and policy coherence. It also promoted some of the key elements that NGOs had campaigned for in 1997 for the panel on new financial mechanisms such as:

- mobilizing domestic financial resources for development;
- mobilizing international resources for development: foreign direct investment and other private flows; and
- enhancing the coherence and consistency of the international monetary, financial and trading systems in support of development.

The EU had committed at the Mexico Conference to increase development aid within the Union to 0.39 per cent by 2006 on the way to 0.7 per cent.

Integrated management and ecosystem approach

> Managing the natural resources base in a sustainable and integrated manner is essential for sustainable development.
>
> (United Nations, 2002c: para 24)

It was clear that the hope of breaking down barriers of silo thinking had not been achieved in the ten years since Rio. As a document the Johannesburg Plan of Implentation did not provide an integrated approach to policy advice.

There was an attempt by the UN Secretary-General to achieve an integrated approach through what he called the WEHAB agenda, covering water, energy, health, agriculture and biodiversity. The papers advocating the strategy were excellent. Unfortunately, he floated the WEHAB agenda between the third and fourth preparatory meetings. It was very far into the process, and too late for the system to refocus itself and the summit to the agenda.

The Secretary-General was late in engaging in the WSSD process, in part because of advice from his office; this hesitation was fatal to getting WEHAB accepted.

Horizontal coordination and cooperation

An area that the WSSD failed to address was the need for a more effective coordination strategy within the UN system working on sustainable development.

In 1992 the Inter-Agency Committee on Sustainable Development (IACSD) was set up to coordinate the UN agencies and programmes responsible for implementing Agenda 21. Thought to be a success when reviewed for the Five-Year Review of Agenda 21 in 1997, it was closed down in the UN reform of the late 1990s. Its last meeting was ironically just prior to the WSSD. This deficiency became more and more obvious after the summit and was still costing gains in effectiveness by resource-poor agencies almost ten years later – raising hopes that governments could be prepared to rectify it at Rio+20.

Vertical coherence

The JPoI looked to strengthen sustainable development governance at all levels and to enhance vertical coherence. It saw the UN General Assembly as a key overseer of mainstreaming sustainable development within the UN system. It looked at ECOSOC to play an enhanced role and for the CSD to develop through a new multi-year programme of activities built on a two year cycle with the first year reporting what was happening on each issue and informing the policy year so that policy redirection could take place to address implementation problems.

For the first time the UN Regional Commissions were given a role in reviewing implementation, which it was hoped would strengthen their work on sustainable development.

External coherence – regional governance

Outside the UN governance, actors were also trying to build more coherence at Johannesburg:

> We live in an increasingly interconnected, interdependent world. The local and the global are intertwined.
>
> (Local Government Declaration to the World Summit on Sustainable Development, 2002)

A significant section of Agenda 21 had been on local authorities, calling for them to develop Local Agenda 21s (LA21s) in consultation with their population. By 2002, there were an impressive total of over 6,000 LA21s in the world, and growing. WSSD tried to build on this by suggesting that there needed to be an

> Enhanced role and capacity of local authorities as well as stakeholders in implementing Agenda 21 and the outcomes of the summit and in strengthening the continuing support for Local Agenda 21 programmes.
>
> (United Nations, 2002c)

A significant non-WSSD outcome from Johannesburg therefore was the establishment of the first sub-national government international network for sustainable development. The Network for Regional Government for Sustainable Development (NRG4SD) brought together 23 regional governments from regions such as Wales, the Basque Country, Bavaria, Gauteng and Rio de Janeiro. They produced the Gauteng Declaration in which they committed them ready to participate:

> Regional Governments both want and need to work with all other spheres of government and with other stakeholders in promoting sustainable development … We consider Regional Governments, from the point of view of proximity, efficiency and spatial dimension, to be strategically located as a necessary and crucial sphere of government for the development of policy for and implementation of sustainable development.
>
> (NRG4SD, 2002)

By the time of WSSD in 2002, there had been over 100 national councils or their equivalent around the world. These were established through a number of diverse mechanisms, such as:

- Argentina and Vietnam: presidential decree;
- Niger and Barbados: ministerial decree;
- Finland: Council of State decision;
- Mexico and Philippines: a law;
- Norway: a letter from the environment minister;
- Ukraine and Grenada: a cabinet resolution.

The structures of national councils for sustainable development differed from country to country but had common characteristics:

- consensus building;
- engagement and partnership;
- fair process; and
- transparency.

(IIED, 2002)

Over the years, these NCSDs have achieved a number of successes. They have proven to be a very effective way for governments to consult with stakeholders and sectors of society. By doing so, they have helped to build support for potentially difficult legislation to move through parliaments.

There were now in place institutions that could help build vertical integration between local communities and cities, regional governments and national sustainable development strategies. The question after WSSD was: would governments at all levels work to strengthen − or weaken − sustainable development policy?

Good governance

The Rio principles had been a high point in codifying profound and far-reaching principles of governance. By the time of WSSD, however, there was an attempt by some governments to try to renegotiate them.

The two most heavily under attack were: the principle of common but differentiated responsibility by some developed countries and the ecosystem approach. Even the previously unquestioned precautionary approach suddenly had to be dropped from references to the biodiversity convention.

The argument was that following those principles cost their industries sales or profits − clearly a priority to them over poverty alleviation or environmental conservation.

It was resolved through a general restatement of commitments to the Rio Principles, but without addressing how the principles should be applied.

The JPoI did state that good governance should be based on:

- sound environmental, social and economic policies;
- democratic institutions that are responsive to the needs of the people;
- the rule of law;

- anti-corruption measures;
- gender equality; and
- an enabling environment for investment.

(United Nations, 2002c: para 4)

The JPoI also commits governments to ensuring the completion of the UN Convention against Corruption. Supported by conservative governments who perceived it as friendly to multi-national businesses, the convention was adopted extremely quickly, in December 2004.

On the other side of the corporate political spectrum, a campaign organized by Friends of the Earth tried to secure a convention on transnational corporations, to regulate business practices. Although unsuccessful at the WSSD, the campaign did lay the foundation for ISO26000, completed in 2010 under the International Standards Organization and which focused on social responsibility.

Partnerships

A major innovation to the implementation model agreed at the WSSD was a vast emphasis on 'partnerships'. While these could include all manner of configurations among governments, businesses, labour organizations, NGOs, and local authorities, the unstated assumption was that they'd be built around the support – and funding – of corporations. Partnerships were originally promoted as a mechanism to support the implementation of a strong JPoI.

The initial push came from a workshop in May 2001 between stakeholders, governments, and UN agencies and programmes. The negotiating process developed a structure for the WSSD that would include two categories of agreements, the traditional 'Type 1' agreements (negotiated documents with principles, commitments and targets by governments) and 'Type 2' agreements (commitments, targets and partnerships by individual or groups of governments and stakeholders).

Through the following eighteen months, G77 and China, and some of the major NGOs, raised concerns that partnerships were going to become a replacement for real political commitments. The US was one of the leading promoters of Type 2 partnerships. The Bush administration seemed to see it as a strategy that fitted its ideological criteria, and could accept – even champion – it as a way to promote the role of the private sector and demote the role of government. Perhaps they also saw it as a way of helping American businesses gain access to investments in attractive developing regions. Others governments supported it simply because it provided a pragmatic way for financing and implementing present agreements at a time when traditional financial resources were increasingly low.

Throughout the process South Africa tried to bridge the two. Guidelines were drawn up by the Bureau to facilitate the link between the two. They suggested that Type 2 outcomes should:

- achieve further implementation of Agenda 21 and the MDGs;
- complement globally agreed Type 1 outcomes and substitute for government commitments;
- be participatory, with ownership shared among partners;
- be new initiatives, or demonstrate added value in the context of the summit;
- be international – global, regional or sub-regional in scope and reach;
- integrate economic, social and environmental dimensions of sustainable development;
- have clear objectives and set specific targets and time frames for their achievement; and
- have a system of accountability, including arrangements for monitoring progress.

Some NGOs were also taken with the idea – seduced, as were governments, perhaps, by the vision of vast untapped pools of corporate wealth just waiting for investment in sustainability projects. Other NGOs were highly suspicious of the structure and the motivation of governments who backed the model. In a consultation in late July 2002 in New York with UN departments and governments, they specified the minimally necessary criteria for any officially approved cooperation. Those included:

- clearly stated targets and timetables for the project;
- full transparency of the planning and finances;
- clear and balanced governance structures, so all partners maintained control;
- sufficient capacity for local partners to follow and respond effectively to the actions of others;
- sufficient resources for local partners to prepare for, consult around, and attend all project meetings; and
- access to information for outside groups to monitor the finances and activities.

None of those were formally agreed within the negotiations, but a set of Guiding Principles for Partnerships for Sustainable Development was annexed to the final report of the summit.

However, the lack of virtually any funding for partners saw the process implode within five years of the WSSD.

Stakeholders

In many people's minds the growth of stakeholder engagement through the 1990s came from the opening up of the process around the Rio Conference in 1992. The nine chapters on the roles and responsibilities of those nine Major Groups had been inspirational. In the five years since the Rio+5 UN General Assembly Special Session the global engagement process itself had around the CSD opened up considerably. The introduction of the stakeholder dialogues had enabled a real political space for discussion with governments to be developed.

The growth in number of international NGOs had been enormous over the ten years. By 2000 there were 37,000 NGOs and it was estimated by UNDP that 20 per cent had been formed in the 1990s.

The preparatory process for WSSD now started to weave the involvement of stakeholders into the preparatory process. It was meant to be a bottom-up process, to enable a review of implementation, to enlighten the preparatory meetings at the regional and then global level. Governments were to identify obstacles in the implantation of Agenda 21 over the past ten years.

Unfortunately, the timelines did not work to enable this to happen effectively. Most countries did not hold consultations in time for them to feed into the process. Even for those countries that did, didn't do it in a way that made it easy to compare the responses and draw conclusions. As a result there was no real regional analysis based on effective input from governments or stakeholders. If the multi-stakeholder dialogues had shown anything, it was that time to prepare was critical.

For the global process, multi-stakeholder dialogues were planned for the second (January) and fourth (June) preparatory meetings and were not focused around the current negotiating text. It was as if the process was designed to focus the stakeholders away from the negotiations. One major NGO who supports dialogues, and had managed the well-respected Bonn Water 2001 multi-stakeholder dialogues, refused to participate and instead focused their efforts on talking to governments directly.

Negotiations

There had been some progress through the preparatory process in agreeing or reaffirming targets for sustainable development. Perhaps the most significant target was the one on sanitation, which subsequently became a part of MDG7 on environment.

The preparatory meetings were tough, especially with Monterrey and Doha casting long shadows on what could and could not be negotiated in the areas of finance and trade.

Resistance from the bloc of politically conservative governments was fierce. Opposition to there being *any* text on finance came from Australia, Canada, Japan and the United States who claimed that 'other fora were and should be dealing with these.'

By the opening of the summit, this 'backlash bloc' of governments had gone so far as to urge language that would have legally defined that any provisional environmental or sustainable development agreement would be subordinate to any conflicting provision of a WTO decision or trade convention. This would have lost serious ground for sustainable development, not only at Johannesburg, but in all processes for years into the future.

For opposite political reasons, France objected to any language that abolished subsidies. This represented the problem of not looking at the economy as the centre of discussion and building the conversation on sustainable development around that. It represented a real failure in 2002.

One participant, Margaret Brusaco-Mackenzie, the former chief negotiator of the European Commission, tried to shift the discussion back to the need to focus on sustainability by emphasizing the security risks of not implementing environmental conservation guidelines.

She raised the issue of environmental security as the next critical issue.

> The connection between environment and security no longer needs to be spelled out. The decline of our environment will lead to war, local and regional, in the medium term and in some cases in the near future (for example, about water rights) unless action is urgently taken … the donor community must think much more holistically. It must incorporate environment and security within its foreign policies; and not only are more funds needed, they need to be spent more effectively and in a coordinated way.
>
> (Brusasco-MacKenzie, 2000a)

Sweden tried to get the issue on the WSSD agenda but failed. It has become a significant part of the narrative of post 2002 sustainable development discourse.

But through the 15 months of negotiations, the end result compared with what had been achieved in Rio was one of significant disappointment. The movement as a whole left Johannesburg bruised and fragmented. It might take another Earth Summit in 2012 for it to come back together.

Johannesburg's Ubuntu

> One of the sayings in our country is Ubuntu – the essence of being human. Ubuntu speaks particularly about the fact that you can't exist as a human being in isolation. It speaks about our interconnectedness. You can't be human all by yourself, and when you have this quality – Ubuntu – you are known for your generosity.
>
> We think of ourselves far too frequently as just individuals, separated from one another, whereas you are connected and what you do affects the whole world. When you do well, it spreads out; it is for the whole of humanity.
>
> (Tutu, 2008)

The South African people and government were amazingly welcoming to the participants in WSSD. As with Rio there was an enormous diversity of events that grew up around the official conference venue, some more focused on the outcomes from WSSD while others used the opportunity of the summit to bring together their members to set priorities for the next decade.

The official Ubuntu village and exhibition was more like the space of the global forum in Rio in 1992 with events, discussions, parties and stalls. Though it was organized by the South African Government, it didn't draw everyone into that space.

Box 7.3 **Key outcomes from WSSD**

- Poverty eradication
  - Poverty: halve, by the year 2015, the proportion of the world's people whose income is less than $1 a day (JPol II.7.a) [1]
  - Hunger: halve, by the year 2015, the proportion of people who suffer from hunger (JPol IV.40.a) [1]
  - Slums: by 2020, achieve a significant improvement in the lives of at least 100 million slum dwellers, as proposed in the 'cities without slums' initiative (JPol II.11) [1]
  - Poverty fund: establish a World Solidarity Fund to eradicate poverty and to promote social and human development in developing countries (JPol II.7.b)
- Water and Sanitation
  - Drinking water: halve, by the year 2015, the proportion of people without access to safe drinking water (JPol II.8 and IV.25.a) [1]
  - Sanitation: halve, by the year 2015, the proportion of people who do not have access to basic sanitation (JPol IV.25.a) [2]
  - Integrated management: develop integrated water resources management and water efficiency plans by 2005 (JPol IV.26) [2]
- Sustainable production and consumption
  - Production and consumption: encourage and promote the development of a 10-year framework of programmes to accelerate the shift towards sustainable consumption and production (JPol III.15)
- Energy
  - Access: improve access to reliable, affordable, economically viable, socially acceptable and environmentally sound energy services and resources, sufficient to achieve the Millennium Development Goals, including the goal of halving the proportion of people in poverty by 2015 (JPol II.9) [2]
  - Supply: diversify energy supply and substantially increase the global share or renewable energy in order to increase the renewable contribution to total energy supply (JPol III.20.e)
  - Markets: remove market distortions, including restructuring of taxes and phasing out harmful substances; improve functioning, transparency and information about energy markets with respect to both supply and demand, with the aim of achieving greater stability and ensuring consumer access to energy services (JPol III.20.p)
  - Efficiency: establish domestic programmes for energy efficiency with the support of the international community; accelerate the development and dissemination of energy efficiency and energy conservation technologies, including the promotion of research and development (JPol III. 20.b, c, h, I, k and III.21)
- Chemicals
  - Health: aim, by 2020, to use and produce chemicals in ways that do not lead to significant adverse effects on human health and the environment (JPol III.23) [2]

continued…

Box 7.3 continued

- International agreements: promote the ratification and implementation of relevant international instruments on chemicals and hazardous waste, including the Rotterdam Convention so that it can enter into force by 2003 and the Stockholm Convention so that it can enter into force by 2004 (JPoI III.23.a) [2]
- Management: further develop a strategic approach to international chemicals management, based on the Bahia Declaration and Priorities for Action beyond 2000, by 2005 (JPoI III.23.b) [2]
- Classification: encourage countries to implement the new globally harmonized system for the classification and labelling of chemicals as soon as possible, with a view to having the system fully operational by 2008 (JPoI III.23.c) [2]
- Renew commitment to the sound management of chemicals and of hazardous wastes throughout their life cycle (JPoI III.23)
- Oceans and fisheries
  - Ecosystem approach: encourage the application by 2010 of the ecosystem approach for the sustainable development of the oceans (JPoI IV.30.d) [2]
  - Fish stocks: on an urgent basis and where possible by 2015, maintain or restore depleted fish stocks to levels that can produce the maximum sustainable yield (JPoI IV.31.a) [2]
  - Fishing: put into effect the FAO international plans of action by the agreed dates for the management of fishing capacity by 2005; and to prevent, deter and eliminate illegal, unreported and unregulated fishing by 2004 (JPoI IV.31.d)
  - Tools: develop and facilitate the use of diverse approaches and tools, including the ecosystem approach, the elimination of destructive fishing practices, the establishment of marine protected areas consistent with international law and based on scientific information, including representative networks by 2012 (JPoI IV.32.c) [2]
  - Reporting: establish by 2004 a regular process under the United Nations for global reporting and assessment of the state of the marine environment (JPoI IV.36.b) [2]
  - Subsidies: eliminate subsidies that contribute to illegal, unreported and unregulated fishing and to overcapacity (JPoI IV.31.f)
  - Governance: establish an effective, transparent and regular inter-agency coordination mechanism on ocean and coastal issues within the UN system (JPoI IV.30.c)
- Atmosphere
  - Ozone: facilitate implementation of the Montreal Protocol on Substances that Deplete the Ozone Layer by ensuring adequate replenishment of its fund by 2003/2005 (JPoI IV.39.b) [2]
  - Access to alternatives to ozone-depleting substances: improve access by developing countries to alternatives to ozone-depleting substances by 2010, and assist them in complying with the phase-out schedule under the Montreal Protocol (JPoI IV.39.d) [2]

- Biodiversity
  - Biodiversity loss: achieve, by 2010, a significant reduction in the current rate of loss of biological diversity (JPol IV.40)
- Forests
  - Assessment: intensify efforts on reporting to the United Nations Forum on Forests, to contribute to an assessment of progress in 2005 (JPol IV.45.b) [2]
  - Action: accelerate implementation of the IPF/IFF proposals for action by countries and by the Collaborative Partnership on Forests (JPol IV.45.g)
- Corporate responsibility
  - Promotion: actively promote corporate responsibility and accountability, including, through the development and implementation of international agreements and measures, international initiatives and public–private partnerships, and appropriate national regulations (JPol V.49)
- Health
  - Health literacy: enhance health education with the objective of achieving improved health literacy on a global basis by 2010 (JPol VI.54.e) [2]
  - Mortality of children: develop programmes and initiatives to reduce, by 2015, mortality rates for infants and children under 5 by two-thirds, and maternal mortality rates by three-quarters of the prevailing rate in 2000 (JPol VI.54.f) [1]
  - HIV/AIDS: reduce HIV prevalence among young men and women aged 15–24 by 25 per cent in the most affected countries by 2005 and globally by 2010, as well as combat malaria, tuberculosis and other diseases (JPol VI.55) [1]
- Sustainable development of small island developing states
  - Global Programme of Action: undertake initiatives by 2004 aimed at implementing the Global Programme of Action for the Protection of the Marine Environment from Land-based Activities to reduce, prevent and control waste and pollution and their health-related impacts (JPol VII.58.e) [2]
  - Tourism: develop community-based initiatives on sustainable tourism by 2004 (JPol VII.58.g) [2]
  - Energy: support the availability of adequate, affordable and environmentally sound energy services for the sustainable development of small island developing states, including through strengthening efforts on energy supply and services by 2004 (JPol VII.59.a [2]
  - Barbados Programme: review implementation of the Barbados Programme of Action for the Sustainable Development of Small Island Developing States in 2004 (JPol VII.61) [2]
- Sustainable development for Africa
  - Agriculture: improve sustainable agricultural productivity and food security in accordance with the Millennium Development Goals, in particular to halve by 2015 the proportion of people who suffer from hunger (JPol VIII.67) [2]
  - Food security: support African countries in developing and implementing food security strategies by 2005 (JPol VIII.67.a) [2]

continued...

Box 7.3 continued

> - Energy: support Africa's efforts to implement New Partnership for Africa's Development (NEPAD) objectives on energy, which seek to secure access for at least 35 per cent of the African population within 20 years, especially in rural areas (JPol VIII.67.j.i) [2]
> - Means of implementation
>   - Education: ensure that, by 2015, all children will be able to complete a full course of primary schooling and that girls and boys will have equal access to all levels of education relevant to national needs (JPol VIII.62.e, X.116.a and X.117.g) [1]
>   - Gender equity: eliminate gender disparity in primary and secondary education by 2005 (JPol X.120) [4]
>   - Education for sustainable development: recommend to the UN General Assembly that it consider adopting a decade of education for sustainable development, starting in 2005 (JPol X.124.a) [2]
> - Institutional framework
>   - National strategies: take immediate steps to make progress in the formulation and elaboration of national strategies for sustainable development and begin their implementation by 2005 (JPol XI.162) [3]
>   - Governance: adopt new measures to strengthen institutional arrangements for sustainable development at international, regional, national and local levels (JPol XI.139)
>   - CSD: enhance the role of the Commission on Sustainable Development, including through reviewing and monitoring progress in implementation of Agenda 21 and fostering coherence of implementation, initiatives and partnerships (JPol XI.145)
>   - Integrated approach: facilitate the integration of environmental, social, and economic dimensions of sustainable development into the work programmes of the UN regional commissions (JPol XI.160, 160.a–d)

Key

[1] Reaffirmation of Millennium Development Goals
[2] New target
[3] Reaffirmation of Rio target
[4] Reaffirmation of Dakar Framework for Action on Education for All

Adapted from UN DESA (2002) and United Nations (2002c)

IUCN and WWF hosted their own miniature global forum at the Ned Bank headquaters – one of the strengths of that venue was it was close to the negotiations and therefore saw a considerable number of delegates attend events there, including: the Green Web in Action exhibition; daily high-level Futures Dialogues led by key figures; and an Investment Fair Kiosk, prepared by Projects Africa.

Some of the other key events and initiatives around WSSD were:

Access Initiative – *World Resources Institute*

The Access Initiative (TAI) is a global coalition of public interest groups collaborating to promote national-level implementation of commitments to access to information, participation, and justice in environmental decision-making.

Aspirations and Reality: Building Sustainability – *Royal Institute of Chartered Surveyors Foundation*

The conference was open to 500 delegates only – it centred on the presentation of the world's most successful examples of sustainable development across the sectors to date. Providing access to world-leading expertise and guidance, this is a valuable learning and networking opportunity for those committed to corporate social, economic and environmental responsibility.

Business Partnership Initiatives – *Business Action for Sustainable Development*

BASD presented 'Business Partnership Initiatives', examples of business working openly with others towards sustainable development.

The Earth Charter Initiative – *Earth Council*

The Earth Charter is a vision for the future that can renew governments' and people's commitment to achieving what began at the Rio Earth Summit, but has lacked a unified framework and solidarity in effort.

Earth Summit 2002 Awards – *Stakeholder Forum and RSA*

The 'Earth Summit 2002 Awards' aimed to encourage further implementation of sustainable development through recognizing, rewarding and publicising 10 years of global stakeholder best practice, which has inspired and will continue to inspire others to work towards the ideals of Agenda 21, as set out at the Rio Earth Summit in 1992.

Earth Summit Campaign – *Globe Southern Africa*

GLOBE Southern Africa's Earth Summit Campaign had four main objectives:

- to mobilize parliamentarians;
- to raise awareness about the summit in legislatures around the world;
- to enable MPs to play a meaningful and active role in the preparations for the summit and at the summit itself; and
- to provide legislators with information and legislative tools they can use after the summit in their respective parliaments.

EnviroLaw Conference 2002 – *EnviroLaw Solutions*

EnviroLaw 2002 examined and explored the negotiation, agreement and ratification of conventions, the application of laws and regulations and their impact on sustainable development.

The Equator Initiative – *UNDP, IUCN, TVE (Television Trust for the Environment)*

The Equator Initiative was designed to reduce poverty through the conservation and sustainable use of biodiversity in the Equatorial belt by fostering, supporting and strengthening community partnerships through the recognition of local achievements, the fostering of South–South capacity building, and by contributing to the generation and sharing of knowledge. The first 'Innovative Partnership Awards for Sustainable Development in Tropical Ecosystems' was presented at WSSD.

Implementation Conference – *Stakeholder Forum for a Sustainable Future*

The Earth Summit 2002 in South Africa launched 23 Type 2 partnerships and allowed stakeholders to come together and work out how to do their part in implementing the Sustainable Development Agreements.

Johannesburg Climate Legacy – *Various Supporters*

Under the umbrella of the WSSD, The Johannesburg Climate Legacy (JCL) measured the CO_2 emissions of the summit (from aircraft flights to electricity used at the event itself). These emissions were offset through investments in carbon-reducing sustainable projects across South Africa. Companies, individuals, governments did sponsor some of this.

Local Action Moves the World – *International Council for Local Environmental Initiatives*

Local Action Moves the World provided an opportunity for local government leaders and their partners to present the key messages from the Local Government Dialogue Paper, the official representation of the local government position, to the summit and the world.

Network for Regional Government for Sustainable Development – Stakeholder Forum for a Sustainable Future and a number of regional governments

WSSD launched the first global network for regional and sub-national level government to work on sustainable development. They agreed the institution and a Gauteng Declaration on their commitments for implementing Agenda 21 and the JPoI.

Responsible Tourism in Destinations – *Stakeholder Forum for a Sustainable Future and UNWTO, Cape Town Authority and various other hosts*

This conference in Cape Town focused on maximizing socio-economic benefits for local communities from tourism ventures, while maintaining the quality of the environment is a major challenge facing all the stakeholders in the tourism industry.

Summit Institute for Sustainable Development – *The Smithsonian Institution*

The SISD was a novel initiative, mobilizing existing WSSD participants to present formal mini-courses on key sustainable development issues that will provide background, tools, and approaches for implementation of sustainable development policies.

The Virtual Exhibition – *Multiple Hosts*

A virtual platform to share sustainable development projects with the world.

WaterDome – *International Water Management Institute, African Water Task Force*

The WaterDome was the main venue during the World Summit where water-stakeholders from public and private organizations got the opportunity to launch and exhibit their activities, policies, initiatives, new technologies and products.

The World Forum for Sustainable Development – *International Research Foundation for Development*

The world forum of the International Research Foundation for Development focused on the implementation of Agenda 21 adopted during the Rio Conference on Environment. Attending were researchers, policy-makers, members of governments and advocacy groups, and all civil society members interested and concerned about sustainable development issues.

The World Sustainability Hearings – *Leadership for Environment And Development*

In an effort to increase effective participation of ordinary people in global governance, the World Sustainability Hearings project was supported by over 40 other civil society organizations who teamed up to provide a stage for their testimony at the Johannesburg summit.

Failed to deliver!

For some organizations that had been working on WSSD for nearly five years, the end result was a huge disappointment. A look back at some of the hopes of early 2000 for what the summit might achieve, compared with what was achieved, does make for sober reading.

Were any of the Osborn Objectives – the key goals for the summit – met?

The first of a revitalized and integrated system for sustainable development would lie in ruins within three years of WSSD, the CSD failed for the first time ever to agree any text, and this was on one of the most important areas: energy.

The second objective, a new deal on finance, had started well only to be completely destroyed by 9/11, and any extra funds that might have been used for sustainable development were hovered up by donor governments to spend on addressing terrorism and fighting two wars.

The third objective was a clear understanding of how governments should move forward nationally on implementing Agenda 21. Within two years of WSSD, governments were closing down their national councils for sustainable development, and within four years people were focusing on climate change and its mitigation and adaptation issues. Some development ministries even stopped coming to the CSD.

The fourth objective that there should be an 'integration of trade and sustainable development' got a sharp no from key governments.

The fifth objective 'support for the Earth Charter as an ethical framework' briefly appeared in the political declaration but only to be stamped on by governments.

The final objective had been a 'clear set of commitments to implement agreed actions'. Here there was perhaps some minor success with the target set for Freshwater. However, even that small hope would be dashed within three years.

As participants streamed out of Johannesburg, far more exhausted and dejected than they had arrived, they heard that the first chair of the Commission on Sustainable Development in 2003 would be South Africa and its excellent environment minister, Vali Mosa. This left the possibility that there might be a chance to follow up, to refocus and agree a work programme that might, just might, offer a second chance. South African President Thabo Mbeki's summed up the hope:

> The peoples of the world expect that this World Summit will live up to its promise of being a fitting culmination of a decade of hope, by adopting a practical programme for the translation of the dream of sustainable development into reality and bringing a new global society that is caring and humane.
>
> (Mbeki, 2002)

These words were to prove hollow within four years of WSSD.

Chapter 8

Chasing dreams

Even before the Johannesburg Summit had ended, serious concerns were being expressed about the future of the CSD and the entire process of implementation. Some of those were raised in an NGO-sponsored event that included representatives of many governments, UN agencies and programmes. Five former CSD chairs spoke: Henrique Cavalcanti (1995), Mostapha Tolba (1997), Simon Upton (1999), Juan Mayer (2000) and Bedřich Moldan (2001). Their analysis of the CSD was very bleak, but they gave strong recommendations for an agenda for the future CSD that would 'monitor, review and implement', and for the need to a focus on a more manageable agenda.

In a paper written just before the WSSD, Simon Upton tried to address why it is that, with all the evidence available, governments were still not willing to act:

> Is it that, deep down, Ministers and their advisers don't really believe there are problems of the sort outlined in this article, but that they're stuck with what they said at Rio and can't easily bail out?
>
> Or is it that countries acknowledge the difficulties but find themselves stuck in a Prisoner's Dilemma in which, lacking the international co-ordination mechanisms, they can only move at the pace of the slowest and most compromised participants?
>
> Or is it all much more innocent – that we simply lack reliable measures of sustainability and a precise definition of the thresholds we dare not cross?

He went on to say:

> Politicians must now decide whether to treat sustainable development as a Holy Grail that is so complex no-one can grasp it ... or to settle on a few concrete problems that can head off human and environmental pressures that are well described but unlikely to evolve tidily according to a timetable or with predictable consequences. If these problems do get out of control, we can be sure that it will be all hands to the pump. But fighting fires is no substitute for investing in prudent, well-researched insurance.
>
> (Upton, 2001)

It was clear from the analysis going into WSSD that the previous ten years had not delivered what had been promised. But would the next ten be any better?

Already the main institution responsible for monitoring the implementation was considered by its previous chairs as lacking. The text in the Johannesburg Plan of Implementation (JPoI) agreed with this perception:

> the Commission needs to be strengthened ...An enhanced role of the Commission should include reviewing and monitoring progress in the implementation of Agenda 21 and fostering coherence of implementation, initiatives and partnerships.
>
> (United Nations, 2002c: para 145)

When governmental bodies are told they need to be strengthened, it's typically diplomatic-speak for saying, 'you haven't done your job.'

So the organizational session of the CSD in 2003 under the chairship of Vali Mosa would play a critical role in determining the approach to sustainable development at the global level for a decade.

Eleventh Session of the Commission on Sustainable Development (2003)

The WSSD had agreed to a system of two-year cycles with negotiations on policy taking place in the second year. The review in the first year was to identify roadblocks to implementation and to suggest policy options to address those roadblocks, as well as to bring 'success stories' to the CSD which could be analysed and replicated.

The CSD agreed a work programme that would accommodate a Johannesburg+15 or a Rio+25, but not a review in five or ten years. This reflected the exhausted and bruised state of governments' reactions after WSSD. They did leave themselves an option to change their minds in 2010, something that would be activated due to the strong leadership of Brazil.

So the outcome from CSD 2003 was that this long 15 years of work now focused around implementation more than policy development. It reflected very much a victory for the United States, supported naively by some NGOs who had bought in to the idea of partnerships being the next 'big idea' to rescue sustainable development.

The 2003 CSD in its organizational plan replaced 12 hours of government–multi-stakeholder dialogue with a return to the pre-1997 approach of letting stakeholders speak at the end of each daily session, if there was enough time. It was described as an opening to more substantive contributions by stakeholders, but as it played out, it was revealed as a massive reversal in participation. Since this CSD session was not well attended by stakeholders, however, the realization of what had been negotiated only became apparent in the following two years.

The CSD adopted a schedule for a series of two-year action-orientated 'Implementation Cycles' focusing on thematic clusters of issues (Box 8.1).

Box 8.1 **United Nation Commission on Sustainable Development Chairs 2003–2012**

- 2003 Vali Mosa, South Africa, Minister of Environment and Tourism
- 2004 Borge Brende, Norway, Minister of the Environment
- 2005 Ambassador John Ashe, Antigua and Barbuda, Minister of Sustainable Development
- 2006 Aleksi Aleksishvili, Georgia, Minister of Finance
- 2007 H.E. Abdullah bin Hamad Al-Attiyah, Qatar, Minister of Environment
- 2008 Francis D.C. Nhema, Zimbabwe, Minister of Environment
- 2009 Gerda Verburg, Netherlands, Minister of Agriculture, Nature and Food Quality
- 2010 Dr Luis Ferrate, Guatemala, Minister of the Environment
- 2011 László Borbély, Romania, Minister of Environment

- Year 1: A review session to evaluate priority concerns in the implementation of a selected thematic cluster of issues, and to prepare discussion in the policy year.
- Year 2: A policy session to take decisions on practical measures and options to expedite implementation in the selected thematic cluster of issues.

Each cycle would address a thematic cluster of issues and cross-cutting issues. It would look at the constraints and obstacles to implementation and look for opportunities and solutions to address these.

For the new two-year cycle some issues seen as cross-cutting were:

- poverty eradication;
- changing unsustainable patterns of consumption and production;
- protecting and managing the natural resource base of economic and social development;
- sustainable development in a globalizing world;
- health and sustainable development;
- sustainable development of small island developing states;
- sustainable development for Africa;
- other regional initiatives;
- institutional framework for sustainable development;
- gender equity;
- education.

There would be high-level segments each year, organized to give guidance, to allow political oversight of decision making, and to encourage interaction with major groups, agencies, funds, programmes and other organizations within the UN systems, and international finance and trade institutions.

Options for the subject of the thematic clusters of issues would be periodically considered and recommendations made to ECOSOC, in its role in promoting system-wide coordination.

A new role for regional commissions was introduced. They were invited, but not formally mandated, to meet prior to and input to the Secretary-General's State of Implementation Report.

In guiding the implementation of the Multi-Year Programme of Work, it was suggested that:

- the review will be dealt in accordance with relevant provisions of Agenda 21, the Programme for the Further Implementation of Agenda 21, and the JPoI;
- thematic clusters of issues will be addressed in a way that considers all dimensions of sustainable development;
- the selection of clusters of issues does not undermine the importance of *all* issues identified in Agenda 21 and the JPoI;
- Agenda 21 and the JPoI: the implementation process should cover all issues equally;
- means of implementation identified in the JoPI should be considered in every cycle, as well as additional cross-cutting issues and other regional activities, including SIDS, Africa and LDC. Refer to Box 8.1;
- the CSD should focus on issues where it could add value to inter-governmental deliberations on cross-cutting issues;
- the CSD should take into account the outcomes of the work of the Open-Ended Ad Hoc Working Group of the General Assembly on the integrated and coordinated implementation of, and follow-up to, the outcomes of the major UN conferences and summits in the economic and social fields;
- the commission may decide to incorporate new challenges and opportunities related to implementation into its Multi-Year Programme of Work.

Problems with the new mandate

Within the agreement on the organizational structure were some important decisions that were to come back and haunt governments in the coming years.

These included not creating an inter-agency body to coordinate the UN system organizations in monitoring and implementing the agreements from Johannesburg. The IACSD had been closed down in 2001, and so the impact of not having it was not felt by 2003.

The lack of a reaffirmation for some of the key elements of the original mandate of the CSD was perhaps the biggest mistake. The original mandate had a very clear review function in relation to funding implementation. Three paragraphs had been critical to determining who turned up to the CSD in the first ten years:

3(c) To review the progress in the implementation of the commitments set forth in Agenda 21, including those related to the provision of financial resources and transfer of technology;

Box 8.2 **United Nations Commission on Sustainable Development Multi-Year Work Programme**

Thematic clusters

2004/2005
- water
- sanitation
- human settlements

2006/2007
- energy for sustainable development
- industrial development
- air pollution/atmosphere

2008/2009
- agriculture
- rural development
- land
- drought
- desertification
- Africa

2010/2011
- transport
- chemicals
- waste management
- mining
- ten-year framework of programmes on sustainable consumption and production patterns

2012/2013
- forests
- biodiversity
- biotechnology
- tourism
- mountains

2014/2015
- oceans
- marine resources
- small island developing states
- disaster management and vulnerability

2016/2017
- overall appraisal of implementation of Agenda 21, the programme for further implementation of Agenda 21 and the Johannesburg Plan of Implementation

---

3(d) To review and monitor regularly progress towards the United Nations target of 0.7 per cent of the gross national product of developed countries for official development assistance; this review process should systematically combine the monitoring of the implementation of Agenda 21 with the review of financial resources available;

3(e) To review on a regular basis the adequacy of funding and mechanisms, including efforts to reach the objectives agreed in chapter 33 of Agenda 21, including targets where applicable.

(United Nations, 1993)

Although it did not have an immediate impact, over the next five years development ministries increasingly took the CSD off their timetables. The new mandate did not require them to attend and report on funding implementation. In fact the Millennium Development Goals became their main focus.

One of the less obvious problems was hidden in Section 2g of the report of the 2003 CSD. It said:

> the discussions of the intergovernmental preparatory meeting will be based on the outcome of the review session, reports by the Secretary-General as well as other relevant inputs. Based on those discussions, the Chair will prepare a draft negotiating document for consideration at the policy session.
>
> (UNCSD, 2003: 2g)

It was a seemingly innocent role given to the CSD chair. But it put the chair not in a 'facilitating role' but one of control, and left out of the CSD Bureau (the five-person coordinating committee of government delegates elected annually by the respective UN regional groups). A number of NGOs raised concerns and suggested that it be changed to 'Chair and Bureau', but were told by seasoned government negotiators that they 'shouldn't worry about these kinds of things, a chair would always work with their bureau'. Within two years, governments were made aware that was a mistake.

Two other organizational issues were to impact on the new CSD: the decision to have different bureaus for the review year and the policy year added to the disjointed nature of the cycles. And the move of the ministerial session from the beginning of the CSD to the end meant there was little room anymore for ministers to innovate and offer leadership.

Some good news

The new CSD introduced some innovation with the Partnership Fair, where existing or new partnership projects could share their experiences, and increase opportunities that they might be replicated. One such partnership, the Supporting Entrepreneurs for Sustainable Development (SEED) Award, was a good example of how to bring replicable projects to the CSD. It represented a Type 2 partnership initiated by the German Government supported by IUCN, Stakeholder Forum, UNDP and UNEP. The goal of SEED was to support the ability of such entrepreneurs to scale up or replicate their activities in order to increase their contribution to their local economies and communities while promoting sustainable management of natural resources and ecosystems and reducing poverty, marginalization and exclusion. During the 2003 CSD, it launched its first set of winners. It never had much money behind it but is one of the few successes of partnership that is still going after ten years and still contributing to the ideas and good practice.

The second innovation was the Learning Centre. It aimed to facilitate training at a practical level on various aspects of sustainable development. Courses were chosen to impart practical knowledge that would enable participants to implement Agenda 21 and the Plan of Implementation in their home countries. It had been built around an innovative programme put together by the Smithsonian Institute for WSSD. The main problem was that the individuals attending the CSD were in most cases policy analysts and advocates – not the people who would find these courses that useful.

Selecting water as the first major issue to go through the CSD two-year cycle helped build on the momentum that had been created during the run up to Johannesburg, when the issue had success with the sanitation goal agreed at the WSSD.

Twelfth and Thirteenth session of the Commission on Sustainable Development (2004–2005)

The first CSD two-year cycle started with a dynamic bureau and chair – the Norwegian Environment Minister, Borge Brende, who worked to try and frame what was a totally new process.

Regional implementation meetings

As the year leading up to the meeting progressed it became clearer that the whole process did not hold together. In the broad area of sustainable development, there are a plethora of processes and issues which must continually be addressed if nations are to avoid catastrophes and make significant progress.

Different regions face different challenges to implementation, and appropriate solutions can often be found at the regional level to suit regional economic, environmental and social characteristics. Arguably, the regional implementation meetings (RIMs) could be a critical mechanism for feeding-in regional experiences into the CSD review. But there are some lessons to be learned.

The RIMs, to be truly effective, needed to follow a coherent and consistent structure, as did their outcomes. In the absence of a reporting framework, the outcomes of the RIMs became largely incomparable and had little impact.

Having similarly structured reports would allow governments to identify areas of commonality where global solutions need to be found to global problems. In addition to this, the RIMs could be used to identify clear implementation gaps that stem from a lack of capacity. The timing of the RIMs also needed to be thought out – before or after the inter-sessional. The process was not clear, and the way in which it would work over the next year – to ensure that momentum and lessons learned were communicated between CSDs and translated into practical policies and actions – was not determined.

Innovation versus tradition

The first week of CSD-12 could be heralded as a success. There was an exciting energy for sharing experiences and for having an honest and frank discussion on the obstacles, challenges and constraints to implementation. There was an appetite for learning from one another, identifying areas of commonality in implementation experiences, and seeking possible ways forward to overcoming these. The delegates seemed to have suddenly learned to become social creatures – rather than creatures of negotiation.

The level of interaction within the formal review sessions was commendable and the discipline of the session chairs in keeping interventions short was remarkable – no mercy was shown. The Partnership Fair was a success, and the Learning Centre was an illustration of how we all still have a lot to learn.

But the level of enthusiasm dwindled and fatigue increased over the second week, as the approach moved back to a more traditional version of a UN session.

Many delegates left the 2004 CSD pondering what lessons they had learned for the policy year and if the sessions had really added any value to achieving the JPoI, and the MDGs.

There were questions to which no answers seemed available. What would be the role of the inter-sessional meeting next year? How would, or should, it be used regarding the five-year review of the MDGs? Since the issue primarily being discussed at CSD was water and sanitation, a major part of MDG7, should the review of the Millennium Development Goals be at the heart of the CSD deliberations? In fact, should the CSD play the role of PrepCom on MDG7?

What was urgently needed was some coordinated strategic thinking and some coherence, consistency and collaboration within the UN and among governments. To some, CSD 2005 would be the litmus test to the new CSD. Was it central to the governments' thinking on sustainable development or had it been replaced by the MDGs?

Problems starting

The chair of the 2005 CSD was Ambassador John Ashe, from Antigua and Barbuda, who had been one of the authors of the original resolution for the Rio conference, in 1991, and was one of the longest serving ambassadors in New York.

His election was not straightforward. The Major Groups women's caucus had wanted the CSD to select its first woman chair, and lobbied strongly for the Brazil's minister, Marina Silva.

One argument put forward against Ambassador Ashe was that the CSD was always chaired by a minister, not an ambassador – though it wasn't actually correct, as the first CSD was chaired by Ambassador Razali of Malaysia. Within 24 hours of this argument, however, the ambassador had been designated by his capital as the first Minister for Sustainable Development of Antigua and Barbuda.

Financing and water

The CSD opened with some good news. Germany had set a date of 2015 (it had been at 0.24 per cent, and by 2010 it was at 0.38 per cent) to reach 0.7 per cent of GDP for development aid, and joined other countries who had more recently promised to reach 0.7 per cent – the UK by 2013 (2005 it was at 0.37 per cent and by 2010 it was at 0.56 per cent) and France by 2012 (in 2005 it was 0.31 per cent and by 2010 it was 0.45 per cent). Of course, countries like Denmark, Finland, Norway, Sweden and the Netherlands had both achieved – and surpassed – the target, years earlier.

Preparing for the discussion on water, the Stockholm International Water Institute (SIWI) released a report that pointed out poor countries with access to improved water and sanitation services had enjoyed annual average economic growth of 3.7 per cent of GDP, while those without adequate investment saw their GDP grow at just 0.1 per cent annually (SIWI, 2005).

The report pointed out that the investments needed were not excessive, and called for donor governments to fund water and sanitation, to also boost countries' economic growth and contribute to poverty reduction. The report found that in countries where water storage capacity was improved, national economies were more resilient to variability in rainfall and economic growth is boosted. Investing in water was good business: improved water resources management and water supply and sanitation contribute significantly to increased productivity within economic sectors. Economic benefits ranging from US$3 to US$34 per US dollar invested would be gained in the health, agricultural and industrial sectors if the MDGs related to water and sanitation were achieved (SIWI, 2005).

The possibility of linking the funding of water implementation to CSD decisions could have represented a real strengthening of the CSD, but instead donor governments looked to the MDG summit in September to make their announcements, and by implication, weakened the CSD meeting.

Stakeholders

The already shrinking political space for stakeholders was now a major issue. The CSD, which had been *the* beacon for stakeholder engagement in the United Nations, had vastly reduced the space for interaction and in 2005 this was going to further erode. For the five-year review of the MDGs, stakeholders were told they would be given only informal hearings, six weeks before the main event – and no interventions at the MDG Review itself. With the support of all the Major Groups, Stakeholder Forum issued a challenge to governments saying that for the remaining session of that current year's CSD, stakeholders would like the following:

- all stakeholders should be able to comment on the chair's text in the Friday session;
- during the negotiations and depending on time, stakeholders should be able to react to the debate with interventions;
- during the high-level session, stakeholders should be able to fully participate;
- stakeholders should have space to contribute to and make statements in the final CSD session.

And for CSD 2006 and beyond, stakeholders wanted to see:

- the reintroduction of the multi-stakeholders dialogues;
- the chair's summary document from the multi-stakeholder dialogue sessions to be entered to the government consultations and the views to be used by the Bureau as legitimate input to the final document.

(Outreach, 2005)

First cycle draws to a close

The water community had done its best to make a breakthrough at the CSD. With the passing of the 2005 target on integrated water resource management, and it not being anywhere close to delivered, many left without any enthusiasm for returning in 2007 when a review of the outcomes would be held. The review became one of the small victories of that CSD.

The other major issue under discussion at the 2004 and 2005 CSD was human settlement. Because of atrocious scheduling, the CSD was meeting at the same time as the Commission on Human Settlements, in Nairobi, so there was little progress that could be made at the Commission.

Ambassador Ashe's chairing had been a controversy throughout the 2005 CSD meeting. He became the first chair to see it as his responsibility to run the whole show. Bureau members were sidelined, other than to chair sessions. Their frustration could be felt, but there was nothing that could be done. The process was like a slow-motion disaster unfolding. By the end of the first week, governments were in informal discussion about meeting in September, around the MDG review, to see what could be done to refocus the Commission.

Fourteenth and Fifteenth session of the Commission on Sustainable Development (2006–2007)

The new CSD chair, Aleksi Aleksishvili, was the Minister of Finance of Georgia. This was the second time that the Eastern European group selected a finance minister, as they did with Bulgaria in 1996. It was also one of the few times an environment minister wasn't chosen as the chair. But again, a chair from Eastern Europe ran into attendance issues. This time it was that his country was struggling with intense financial problems, and the chair was absent during key meetings.

On the bureau that year was Yvo de Boer, who would later head the UNFCCC secretariat before and during its critical 2009 Copenhagen meeting. He played a highly constructive role on the CSD Bureau from the beginning, in reaching out to stakeholders and governments to try and seek solutions to the problems in the negotiations. He had the trust of many governments.

The CSD had failed to address its input to the MDG review. It could have acted as a prepcom for MDG7 which included the goals on water and sanitation. By not doing so it showed its lack of relevance to development ministries and developing countries as a place for serious financial conversations on implementation.

The future of the CSD: a workshop

A number of the governments who were very concerned about the progress of the first two-year CSD cycle met with stakeholders and UN agencies and programmes in New York in the margins of the MDG review.

At a workshop organized by an NGO, they identified a number of issues that governments should consider planning for the next two-year cycle. The first was the need for a standing agenda item on 'Institutional Framework for Sustainable Development': it would allow the commission to provide guidance to, and receive guidance from, other UN and international decision-making forums.

It was clear from the first cycle that there was a need to balance the normative (the negotiated decision) and non-normative outcomes (the matrix and voluntary commitments) of each session.

The first cycle had resulted in an outcome that contained both normative and non-normative outcomes. While each was important in moving the sustainable development agenda forward, each also had an appropriate place in the process. Governments recognized a need to streamline, not eliminate, the negotiating process. Similarly, a more elaborate preparatory process was needed, where important actors received the necessary support in time to enable them to make an active contribution to the preparatory negotiations.

The CSD had to be able to address politically sensitive issues. The fact that in its first cycle the CSD had avoided discussions on issues that were critical to sustainable development clearly wasn't good enough.

During the first cycle, for example, the almost total lack of discussion on international waters, the role of large-scale water infrastructure, subsidies and trade due to the 'political sensitivity' of such issues demonstrated great weakness of a body like the CSD, and why many actors were starting to move away from it and not see it as 'the authoritative body on sustainable development'. If the CSD was to remain relevant, it must have the political will to discuss sensitive issues and deliver advice. This next cycle on energy, another environmental security issue, would be critical to the CSD's reputation and standing.

The workshop that September reinforced the message that there needed to be momentum continuity between the review and policy years. The challenge for the second multi-year programme of work was to ensure that the debates, discussions and proposals from the review session are adequately translated into meaningful policy recommendations and actions by the intergovernmental preparatory meeting and the policy session. Necessary resources would need to be allocated to ensure continuity. It was also clear that the timetable set by the CSD for national reports, regional meetings and input to the global process just didn't work.

In the previous ten years, governments would finish one CSD in April/early May and then work with their stakeholders on the next year's CSD by preparing input to the UN Task Manager on the relevant chapter of Agenda 21 by the end of October.

Now, to fit into regional meetings scheduled in October to December, national reports needed to be sent in by the end of July. The result was a complete implosion of multi-stakeholder processes at the national level.

The meeting also underlined the role not of the chair but of the bureau in management and coordination. The bureau has long been recognized as a major factor determining the success or failure of the CSD cycle.

The meeting also recognized the need to enhance the contribution of Major Groups. The workshop supported the recommendations made by stakeholders at the end of the previous CSD for the re-introduction of multi-stakeholder dialogues.

Finally the problems with the lack of common approach by regional commissions were addressed and a standardized, but flexible, format for the RIM outcomes should be considered (Sherman et al, 2005).

The UN secretariat was very hostile to the workshop even though it had wide government support (with even the lead US negotiator attending during his family holiday). Only the issue of the dialogues was taken up by them, with no enthusiasm and no proper preparation. The previous 12 hours of dialogue were now reduced to only 90 minutes and just were stilted statements. Within a cycle they disappeared again.

Millennium Development Goals Review (MDG+5)

The MDG+5 review conference didn't just address the MDGs but responded also to the UN Secretary-General's 'High-level Panel on Threats, Challenges and Change'.

The MDG+5 review did set in motion the setting up of two new UN institutions, the Peace Building Commission and the Human Rights Council, taking the place of the discredited Human Rights Commission. It also underlined the UN Delivering as One initiative at the country level and launched a process on reviewing the way the UN dealt with environment.

In the area of new initiatives or new money announced, the clear winner was both surprising and very gratifying for the future. That country was China.

China clearly led the way and its President Hu Jintao made a number of announcements:

1. China has decided to accord zero tariff treatment to certain products from all the 39 LDCs having diplomatic relations with China, covering most of the China-bound exports from these countries.
2. China will further expand its aid programs to the Heavily Indebted Poor Countries (HIPCs) and LDCs and, working through bilateral channels, write off or forgive in other ways, within the next two years, all the overdue parts as of the end of 2004 of the interest-free and low-interest governmental loans owed by HIPCs having diplomatic relations with China.
3. Within the next three years, China will provide US$10 billion in concessional loans and preferential export buyer's credit to developing

countries to improve their infrastructure and promote cooperation between enterprises on both sides.

4. China will, in the next three years, increase its assistance to developing countries, African countries in particular, by providing them with anti-malaria drugs and other medicines, helping them set up and improve medical facilities and training medical staff. Specific programs will be implemented through such mechanism as the Forum on China Africa Cooperation as well as bilateral channels.

5. China will help train 30,000 personnel of various professions for developing countries within the next three years so as to help relevant countries to speed up their human resources development.

(Jintao, 2005)

It would be fair to say that the G8 had made its commitments on debt relief at the Gleneagles meeting in July when the world's richest countries – the G8 – agreed to write off the US$40bn (£22bn; €33bn) debt owed by 18, mainly African, countries.

Sixteenth and Seventeenth Session of the Commission on Sustainable Development (2006–2007)

The 2006 and 2007 CSDs became known as the Energy Cycle, as the previous years had been called the Water Cycle. Each of these issues had no obvious home in the UN system.

On its energy discussion, however, the CSD faced a significant challenge. The previous CSD had been difficult, even with two years of preparation. And the WSSD discussion on energy had not resulted in any target. The OPEC countries had constantly fought against the CSD making any significant advancement on the energy issue, and often succeeded at blocking action. The UNFCCC talks were moving to a new target to succeed Kyoto, and started to attract many NGOs as it gained momentum to set a post-Kyoto target. This left little space or interest from many of the major NGOs to invest effort in the CSD energy cycle – having put considerable work into WSSD without success, they were reluctant to engage in numbers at the CSD.

In an interview at the end of the 2006 CSD, bureau member Yvo de Boer provided insightful analysis about what should happen to the commission:

I would suggest putting the review and policy cycle back to back. The fact that there is a whole year between the analysis and developing policies to address them is a weakness in the current process. For example, here at CSD 14 a significant consensus has emerged on the problems and what needs to be done. We have even heard positive statements from oil producing countries about oil as a problem and the need to move to cleaner technology. It is a pity that we cannot capitalize on that progress right away. I would welcome a conversation on this.

Also, if the CSD is the custodian of the Johannesburg outcome, and if UN system and Bretton Woods institutions are the institutional mechanisms to deliver on that outcome, then I think we should be asking them in a very explicit way to respond to the challenges identified in the analytical cycle. We need a lot more accountability in the exercise of implementing sustainable development commitments.

(de Boer, 2006)

As the closing session of the CSD elected its new chair for the 2007 policy year on energy, participants realized just how challenging it was going to be. The new chair was H.E. Abdullah bin Hamad Al-Attiyah, minister from Qatar. An OPEC member was now going to chair the energy session of the CSD dealing with policy.

Security Council and environmental security

Shortly after the 2006 CSD concluded, the UK government decided to take the issue of energy and climate security to the UN Security Council. As the wars it was fighting in the Middle East continued, the cost and availability of energy had been increasing in impact on economic and politics, and the UK thought it was time to move debate on it to a far more powerful UN body.

Farukh Amil, the deputy permanent representative for Pakistan and chair of the group of 77 developing countries, made some powerful arguments against the inclusion of climate change in the Security Council. While recognizing the importance of the issues for the achievement of sustainable development, developing countries felt addressing the climate and security nexus was the responsibility of the General Assembly and the Economic and Social Council.

Amil argued that climate change already has a binding multilateral agreement, the UN Framework Convention on Climate Change (UNFCCC), as well as the Kyoto Protocol. 'No role was envisaged for the Security Council,' he said.

Stakeholders and climate activists were divided. The international environmental NGO Worldwide Fund for Nature (WWF) called on the Security Council to initiate a global cooperative energy and climate strategy.

Other NGOs opposed the issue being discussed at the Security Council. Supporting G77, they argued, was wrong as it excluded over 180 countries; and of the council's permanent five members, four were developed countries which would be interested in securing their energy and climate requirement. The Security Council decided not to take up the issue, agreeing to send it back to the CSD to deal with in its 2007 policy year.

The end of the CSD?

It is not pleasant to describe the disappointment and despair that many experienced participants felt at the end of the CSD in 2007.

They saw the G77, the voice of the world's developing and poor countries, being held hostage by its oil-exporting members. There were attempts to create a balance among the diverse interests of its members – especially those other, far poorer members like the Alliance of Small Island States (AOSIS), whose economies, culture and very existence were threatened by the actions of the mega-rich oil nations – without any success as the dominance of fossil fuel won the day.

They watched G77 try to add paragraphs in support of nuclear energy towards the end of the negotiations, and some of the developed countries like the US, Canada and France who supported nuclear power quickly suggested additional amendments. Time-bound targets seemed to be out now, only nine months after the MDG review had encouraged them. The US, who had previously championed the MDGs (under President Clinton in 2000) now was arguing that time-bound targets, even voluntary ones, would impact on the sovereignty of countries – an extraordinary comment from a country claiming a right to pre-emptively launch military invasions against other sovereign nations.

They heard Gro Harlem Brundtland, the eminent former chair of the Commission on Environment and Development (who produced the 1987 report *Our Common Future*, otherwise known as the Brundtland Report), say in opening the high-level segment that 'failing is not an option'.

But the final session of the CSD did fail – for the first time – to agree any text at all. After the EU attempted to demand strong targets on energy, and was rebuffed, it refused to accept a watered-down chair's text. There was no time left. It was as if sustainable development had run its course and now this process was effectively dead.

The angry scenes at the end of the CSD were added to as the Africa Group put forward their candidate to chair the next CSD: Francis D.C. Nhema, the Minister of Environment for Zimbabwe. Developed countries tried unsuccessfully to block his election and for the first time in a long time at a CSD, the BBC cameras were waiting outside the negotiating room not to hear about what had happened in the energy cycle disaster, but to interview governments over the election of the Zimbabwean chair of the next CSD. Could the CSD fall any further? Was sustainable development dead?

In the depths of such abysmal failure, it felt like it would take a cosmic shift to pull sustainable development out. Someone had to offer leadership. And in September it became clear where that would come from. At the UN General Assembly opening session, President Luis Inacio Lula da Silva – the hugely popular, previously prosecuted, labour union organizer, who now presided over a booming economy as president of Brazil – proposed organizing a 'Rio + 20' Conference in 2012.

Taking the opportunity to commemorate his country's international success 20 years earlier, and perhaps to congratulate his own achievement in maintaining a thriving prosperity for business while increasing, not decreasing, workers' standards of living and environmental protection, he said:

> I propose that we hold a new conference, in 2012, to be hosted by Brazil: the Rio + 20 Conference. If we want to salvage our common heritage, a new and more balanced distribution of wealth is needed, both internationally and

within each country. Social equity is our best weapon against the planet's degradation.

'It is unacceptable that the cost of the irresponsibility of a privileged few be shouldered by the dispossessed of the earth,' declared the President Lula, calling for industrialized countries to 'set the example', while emphasizing that 'developing countries must also help in combating climate change'.

(da Silva, 2007)

For the UN and governments, it provided a chance to break the limitations of the fifteen-year CSD cycle, reinvigorate environmental negotiations and refocus the sustainable development system.

But there was a muted response from other governments to Brazil's suggestion.

Eighteenth and Nineteenth Session of the Commission on Sustainable Development (2008–2009)

Something had to change within the CSD, and it did when the director of the Division on Sustainable Development left, and was replaced after CSD 2008 by Tariq Banuri, a very well-respected academic and co-winner of the Nobel Peace Prize as a leading contributing author of the IPCC.

Perhaps after such controversy, the Zimbabwean chair ran the 2008 CSD as a team effort with his bureau. The more cooperative approach was what was needed after so many harsh words in 2007.

The main issue for 2008–2009 was agriculture, perhaps as difficult an issue as energy had been. And it would be with a chair whose country was experiencing the worst food crisis since government records began.

Increasingly the term food security was becoming part of the global discourse – linked to the oil price rising towards the highest level ever, US$147.27, on 11 July 2008.

At the same time, the first tremors of the global financial earthquake were starting to hit. In February 2008 the UK government had nationalized Northern Rock, and in March, Bear Stearns collapsed. The financial crisis would peak during September and October 2008.

Against this backdrop, perhaps governments were a little fragile and looking for places where some cooperation could benefit all.

Just before the CSD session in 2008, the UN published two reports (*Trends in Sustainable Development*, and *Trends in Sustainable Development – Africa*) that underlined that efforts to reduce poverty and improve food security were being seriously hampered by the decline in support for agriculture in developing countries.

One of the key findings was that strong economic growth in agriculture is four times more effective in benefiting the poorest half of the population than growth in other sectors. 'These reports clearly show the recent neglect of agriculture in donor and government spending priorities,' said Sha Zukang, UN Under Secretary-General for the Department of Economic and Social Affairs.

Raising agricultural productivity through environmentally sustainable action will be key to feeding a growing world population with rising incomes and changing dietary preferences.

(Zukang, 2008)

One of the main issues put forward to be addressed was that of the need for sustainability criteria for biofuels. In the ministerial dialogue the Irish Environment Minister Gormley reiterated his position on biofuels, saying we need to assess their sustainability, and put criteria in place.

He warned not to be too reliant on technology:

We have to be careful about this emphasis on technology, in particular carbon storage. This is technology which is still in its infancy. I myself feel that if we rely too much on a technological fix, we're going to be waiting around. Technological fixes often allow you to continue business as usual. It's a dangerous game to play. It requires courage and commitment that isn't there right now.

(Gormley, 2008)

He also called for the establishment of an equivalent to the Brundtland Commission to be set up and produce a seminal report for 2012.

The 2008 CSD had started a rebuild. Participants felt the discussion had been good, and even for such a difficult issue as agriculture there had been positive ideas and clear recognition of the need to invest more in agricultural sustainability.

The session provided an opportunity for focusing discussion around the need for infrastructure for sub-Saharan countries, to enable them to better conduct trade. It also showed the inter linkages with the water agenda. With 40 per cent of the Earth's land surface being drylands, and one-third of the world's population living there, (who have the lowest GDP per capital and highest infant mortality rates), it set up the policy year with some high expectations.

It was with enormous relief that when the CSD selected its next year's leadership, it chose its first woman chair in Gerda Verburg, the Netherlands Minister of Agriculture, Nature and Food Quality. The Netherlands had been a real supporter of the CSD from the beginning and if anyone could drag it back to relevance it would be the Dutch.

Rio+20

Brazil had been working through 2008 to pull G77 governments on board to the idea of a new summit in 2012. It finally succeeded.

On October 27, in a statement to the Second Committee of the UN General Assembly, the Group of 77 countries and China expressed their firm support for a Rio+20 Earth Summit in 2012. Highlighting the 'multidimensional' threats the world now faced, and stressing that many of the challenges identified in Rio in 1992 'still remain or have worsened', the G77 and China called for the summit to be hosted by Brazil in 2012 to 'review and assess progress'.

Guided by the 'Rio Spirit', a Rio+20 Summit should provide the necessary political impetus for the range and level of action required to bridge the implementation gap. In this context, the G77 and China supports and welcomes the offer of the Government of Brazil to host such a Summit in 2012.

(G77, 2008)

The group also drew attention to the importance of 'sustainable patterns of production and consumption, with developed countries taking the lead'.

Eight days later, on 4 November, the day of the US presidential election (when US President Barack Obama was elected under a slogan 'yes, we can'), Antigua and Barbuda, acting as chair of the G77, tabled the draft resolution for the 63rd session of the UN General Assembly calling for a World Summit in 2012. The resolution expressed deep concern that 'although some progress has been achieved since the landmark Conference of 1972, the Earth Summit in 1992 and the World Summit on Sustainable Development in 2002, there is a persistent implementation gap, and many commitments by the international community have not been fully met'.

The resolution highlighted the importance of a World Summit in 2012 to 'renew political commitments at the highest level and for an overall review of the implementation of Agenda 21 and of the Johannesburg Plan of Implementation'.

The resolution welcomed and accepted Brazil's offer to host the summit, which would provide the opportunity to 'review and appraise' progress achieved, as well as 'identify further measures' for enhancing implementation of existing commitments (United Nations, 2009).

Donostia

Less than ten days later, on 13 and 14 November 2008, Stakeholder Forum with support of the Basque regional government in Spain hosted an informal multi-stakeholder workshop in San Sebastián to kick-start discussions on the options for an Earth Summit in 2012.

The workshop convened representatives of governments, civil society, intergovernmental organizations and UN agencies to discuss the challenges. Nitin Desai, former Secretary-General of the World Summit on Sustainable Development in 2002 contributed to the discussions via video-link, as did John Scanlon, then Principal Adviser to the Executive Director at UNEP.

All participants expressed overriding support for the hosting of a Rio+20 Earth Summit in 2012, and the discussions and proposals that arose in the meeting were captured in the 'Donostia Declaration' – a document making the case for a summit and providing recommendations as to its focus.

The meeting identified four areas for the summit to address:

- a review of previous commitments;
- emerging issues, e.g. energy security, water security, food security and ecosystem security and their inter-linkages;

- the green economy; and
- sustainable development governance.

The declaration also called for the mobilization of the general public in a solutions agenda and the negotiations of two conventions – one on Principle 10 of the Rio Declaration, covering access to information, public participation and environmental justice, and one on corporate accountability and responsibility, particularly relevant in light of the irresponsible role that the financial institutions had played in creating the economic crisis gripping the world.

Developed countries were not ready for a debate on a new summit, and the UN General Assembly deferred the discussion to the next year.

Commission on Sustainable Development (2009) – It's the economy, stupid!

The 2009 CSD opened with the global financial crisis in full force. Most developed countries had undertaken stimulus packages. Some focused more on greening the economy, the most obvious being South Korea, who used the opportunity to invest in its future with over 80 per cent of their recovery package 'green' according to HSBC. It was estimated it would create over 950,000 new jobs.

Not surprisingly, the UN Secretary-General, Ban Ki-moon, praised the example set by his homeland.

What should these green New Deal programmes consist of? According to Achim Steiner, UNEP Executive Director, they should include:

> Investments in clean-tech and renewable energy; infrastructure such as railways and cycle tracks and nature-based services like river systems and forests, can not only counter recession and unemployment but can also set the stage for more sustainable economic recovery and growth in the 21st century.
>
> (Steiner, 2009)

The second highest investor in a green economy was the European Union, at 59 per cent of the union's recovery budget earmarked for green activities. Third was China (38 per cent) which seemed to be also using the recession to rid themselves of some of their most polluting factories. But where, according to the HSBC report, did that put the countries who were being most vocal about the need for a Green New Deal? Not in a very impressive position – France (21 per cent), US (20 per cent), Germany (13 per cent) and the UK (7 per cent).

Returning to Rio

At the CSD intersessional in February 2009, Stakeholder Forum hosted a dinner attended by 23 individuals from 16 national governments, three non-governmental organizations, one independent consultancy and one university, all talking in their personal capacity under the Chatham House Rule.

Attendance at the informal dinner was from the following governments: Austria, Brazil, Canada, Czech Republic, France, Germany, Italy, Malaysia, Netherlands, Norway, Portugal, South Africa, Sweden, Switzerland, UK and USA (including from the Obama transition team). Discussion focused on a number of areas, including global environmental governance, common but differentiated responsibility, financial implications of a summit, and how to avoid 'summit fatigue'.

Though many governments stated clearly that they had not yet developed positions, the meeting provided an opportunity to exchange ideas on the role a summit could play.

A 'non-paper' was issued based on the discussions at the meeting – the 'non-paper' status indicating that the points outlined did not necessarily represent agreement from all who attended, but rather captured the discussions that took place.

Three aspects seemed positioned to contribute to the success of a summit in 2012:

- the signs that the US, with a new progressive president, was ready to re-engage in multilateral negotiations, especially regarding sustainable development;
- the financial crisis which, even as constituting a major stress for the global economy, calls for re-evaluating the role of the state as regulator;
- the many economic stimulus packages providing the optimal opportunity to launch a 'green economy' and planned growth in the context of sustainable development.

A Rio+20 Summit in 2012 would provide a time-frame for achieving concrete results stemming from specific policies and actions. The meeting would serve, in this regard, as a stimulus for action. The following areas and actions were identified for consideration:

- conduct a review of IEG;
- examine and renew, at the highest level, the long-term political commitments made since the Rio Summit and review the fulfilment of the commitments deriving from the JPoI;
- address the gaps in implementation of Agenda 21 and the Multilateral Environmental Agreements (MEAs).

The non-paper ended with a stark warning:

We are living in worrying times. Human, economic and environmental security are key issues. The Rio Summit managed to generate energy in support of sustainability. If we can tap the [new] emerging spirit of cooperation and action, a whole generation could join in and lend impetus to the effort.

(Stakeholder Forum, 2009)

Economic bubbles

In a slightly amused way, the stakeholder newsletter *Outreach* published an article about economic bubbles and reminded the incoming chair that the Dutch had invented them. It seems that tulips were introduced into Holland in the mid-16th century from the Ottoman Empire. Very soon afterwards there was 'tulip mania' leading to the first economic bubble. At its height, tulip contracts sold for 10 times the annual income of skilled craftsmen. Speculation on the price going up saw masses of people buying or becoming involved in 'bulb futures', which helped to create a tulip bubble.

It was reported that a single bulb of the Viceroy type could be traded for: four fat oxen, two right fat swine, twelve fat sheep, two hogsheads of wine, two lasts of wheat, four lasts of rye, four tuns of beer, two tons of butter, 1,000 pounds of cheese, a suit of clothes, a silver drinking cup and a complete bed.

Eventually this tulip bubble burst when the demand collapsed. People were left holding bulbs that had cost them 10 times the amount the market was now selling them for. The tulip bubble was the first of many, but 500 years later governments had still not prepared for the crisis they were experiencing now.

A success at last

The 2009 CSD was the first under the new director of the DSD secretariat, Tariq Banuri; unlike his predecessor he made himself available for delegates and stakeholders to meet and discuss issues. This made for a much more dynamic and inclusive space in the CSD for all.

The outcome document recognized the need for an integrated approach towards building a new agricultural model which would be people-centred and knowledge-based. The push for a sustainable 'green Africa' to help boost agricultural productivity, food production and national regional food security as well as supporting ecosystem functions was a good development.

On biofuels, the EU announced an agreement among its countries to produce an agreed certification system of biomass production which included social considerations – a very important outcome. With a clear recognition of the role of water in agriculture, the CSD closed on a high.

The chair, in her final day's session, called for agriculture to be included in the climate negotiations. Two years later that too was considered properly by the UNFCCC.

Tariq Banuri put it very well at the end of the session when he said:

Sustainable development is the bridge between the North and the South. It is a means of building trust between the North and South. Building trust is particularly important today, when several crises call for common solutions. The Commission's success is an important expression of common commitment that can enhance such trust, and have positive effect on other actions and negotiations, such as the Copenhagen climate change conference.

(Banuri, 2009)

COP 15 – The Copenhagen Collapse

As the CSD closed in May 2009, preparations for the Copenhagen Climate Summit were starting to gain pace. But the dynamics of the negotiations were pulling in unusually contradictory directions.

Attending Copenhagen were around 45,000 people – equivalent to the Earth Summits – with extraordinary participation from corporations. Except that this was in the midst of an unusually cold and snowy Danish winter – a frozen Earth Summit.

The EU had pledged increasingly greater CO_2 reductions – eventually promising 20 per cent below 1992 levels, by 2020. There was talk it would go to 30 per cent, if the US did so too.

But the US position wasn't clear. President Obama had supported a bill in the US Senate to establish a system of carbon trading and limits. He had also included a significant 'green economy' component in an earlier US economic investment package. But the energy legislation had stalled, and it became clearer that he had used virtually all his political capital to pass health reforms and an economic package. Obama – facing unrelenting partisan obstructionism, and despicable personal attack – seemed not ready to act.

This was confusing to many outside the US, even those working in government. Most didn't seem to realize that the US does not have a unilateral decision-making process. In virtually every other country, when the head of state speaks, he or she acts for the government as a whole. In the US, Congress has the last say.

Meanwhile, politically conservative countries (such as Australia, Canada and Japan) that had previously been strong supporters of Kyoto were now backing away from setting limits. Russia waited, playing its now typical game of 'who will bid the most for my vote?'

The developing countries maintained their position that advanced economic powers had to fulfil their obligation to take the lead in reductions in any new agreement. The oil producers of OPEC as always tried to undermine any agreement, but they were losing much of their influence on the poorer nations. AOSIS had courageously taken a stand as the political and moral conscience of governments.

Reacting to accusations from the American political right that developing countries were in some way gaining an unfair economic advantage (an ironically hilarious analysis), the US and China had been trading accusations and ultimatums – each demanding that the other 'go first' on agreeing to reductions. They seemed to be setting the stage for each blaming the other for a conference failure.

And as the extraordinarily complex negotiations ran into chairing failures, they ground to a halt.

With the summit indeed on the verge of failure, at virtually the last minute Obama unilaterally, and undiplomatically, approached China. And with only hours to go before its close, the US and the BASIC group (Brazil, South Africa, India and China) of rapidly growing economies suddenly emerged with the Copenhagen Accord, a non-binding agreement that only required countries to declare their voluntary contributions to greenhouse gas reductions.

It was a face-saving device, but made nonsense of the previous nine months of intensive negotiations. Stunned delegates reluctantly agreed to 'recognize' the deal, but it was a massive step backward. Momentum on another major intergovernmental environmental process had gone into reverse. As Richard Black the BBC environmental journalist put it:

> After Copenhagen, there is no 'developing world' – there are several. Responding to this new world order is a challenge for campaign groups, as it will be for politicians in the old centres of world power.
>
> (Black, 2009)

Rio + 20

While this was going on, the Second Committee of the UN General Assembly was debating Rio+20. Some of the Copenhagen negotiators came back and then immersed themselves in the final negotiations for Rio+20, which passed as a resolution on Christmas Eve, 2009.

Developed countries reluctantly went along with the decision to organize a conference. Breaking the log jam was France. In a statement issued after a state visit to Brazil on 6–7 September 2009, Nicolas Sarkozy announced France's support for an Earth Summit in 2012:

> France fully supports the desire of Brazil to organize an Earth Summit in 2012, 20 years after the first Summit was held in Rio de Janeiro in 1992. This new Summit will provide a unique opportunity to make decisive and necessary progress in international governance of the environment.
>
> (Sarkozy, 2009)

Within three weeks the European Union endorsed the summit, followed by the US and the other developed countries.

The final outcome was to agree a summit addressing a similar set of priorities to those suggested in 2008 in the NGO-led Donostia Declaration. The UN General Assembly agreed:

> The objective of the Conference will be to secure renewed political commitment for sustainable development, assessing the progress to date and the remaining gaps in the implementation of the outcomes of the major summits on sustainable development and addressing new and emerging challenges; the

focus of the Conference will include the following themes to be discussed and refined during the preparatory process: a green economy in the context of sustainable development and poverty eradication and the institutional framework for sustainable development.

(United Nations, 2009: para 20a)

Twentieth and Twenty-first Session of the Commission on Sustainable Development (2010–2011)

The chair of the 2010 CSD was Dr Luis Ferrate, Minister of Guatemala, a country that over the previous few years had played a very supportive and crucial role within G77. It was also one of the key countries writing the UN GA resolution for Rio+20.

Sustainable consumption and production

The Marrakech Process had been organized in response to the WSSD's call for a ten-year process on sustainable consumption and production (SCP), after a workshop in Marrakech, Morocco, in June 2003. It was set up as a global multi-stakeholder process to support SCP implementation and develop a Global Framework for Action – the so-called 10-Year Framework of Programmes on SCP (10YFP).

The goal was to accelerate the shift towards SCP patterns, de-linking economic growth from environmental degradation. It set up seven task forces to work on:

- Cooperation with Africa;
- Education for Sustainable Consumption;
- Sustainable Buildings and Construction;
- Sustainable Lifestyles;
- Sustainable Products;
- Sustainable Public Procurement; and
- Sustainable Tourism.

The CSD backed the need to have much stronger linkages to eradicating poverty within the 10YFP and that there should be a set of policies and measures developed that would include voluntary, market-based and regulatory. Though it has to be underlined that the 10YFP represented such a low achievement on SCP after nearly 20 years.

Chemicals

Although much of the focus was on SCP there was also solid work being done on chemicals that involved participants from governments and the chemical industry. There was support for the full implementation of existing chemical arrangements such as the Strategic Approach to International Chemicals Management (SAICM) to support stopping the transfer of obsolete technologies to developing countries and to taking action against countries that illegally export chemicals.

NGOs felt that the chemical text had not addressed some significant areas such as:

- capacity building in developing countries to deal with the increased use of chemicals;
- the lack of consumer information on chemical content of products;
- a coherent approach on chemicals in electronic and electrical products waste;
- the lack of a global framework to deal with asbestos.

Mining

A new issue for the CSD arrived with some very strong viewpoints, particularly from the indigenous peoples' community. The chair's text did contain respect for human rights and support for the relevant ILO conventions. It also talked about sharing benefits with local communities and possible no-go areas for mining.

But ultimately mining is unsustainable, a point underlined by a number of the major groups in their interventions.

Lucy Mulenkei from the Indigenous Information Network pointed out that:

> Mining is a grave concern to us because we are disproportionately affected by mining and most subjected to violations of our rights and the environment on which we depend.
>
> (Mulenkei, 2010)

UNCSD2012 – PrepCom I

Earlier in the year, governments had agreed to their bureau members for the new summit. An enlarged bureau of ten was elected with two co-chairs this time – Ambassadors John Ashe and Park In-kook of South Korea. Other bureau members were:

- Ambassador Maged A. Abdelaziz (Egypt);
- Ambassador Charles Thembani Ntwaagae (Botswana);
- Asad M. Khan (Pakistan);
- Tania Valerie Raguž (Croatia);
- Ana Bianchi (Argentina);
- Paolo Soprano (Italy);
- Maria Teresa Mesquita Pessoa (Brazil);
- Jiří Hlaváček (Czech Republic); and
- John M. Matuszak (United States).

It was clear again that G77 took the summit much more seriously than the developed countries, shown by the high-level selection of their bureau members

---

Box 8.3 Summary of government statements on the two themes of Rio+20

Green economy

- Supportive: Barbados, China, EU, Indonesia, Norway, Republic of Korea, Senegal, Sweden, Switzerland, USA, FAO, UNEP, UNIDO, Workers and Trade Unions and some NGOs
- Unclear: Argentina, Australia, Colombia, Ecuador, Egypt, G77 and China, Mauritius, Mexico, Uruguay, Women, NGOs, Local Authorities
- Sceptical/Negative: Bolivia, Brazil, Cuba, Japan, Indigenous Peoples

Institutional framework for sustainable development

- Supportive: Australia, Bolivia, Brazil, Cuba, EU, Egypt, Indonesia, Japan, Mexico, Montenegro, Norway, Switzerland, USA, UNEP
- Unclear: China, Workers and Trade Unions, Science and Technology
- Sceptical/Negative: some NGOs

---

compared with lower-level civil servants from developed country ministries. Also, where were the representatives from finance ministries? Again, developed countries put in environmental civil servants to cover an area that they had no control over.

The 2010 CSD ended with the election of the new chair, László Borbély of Romania, and the first preparatory session for Rio+20 started on the following Monday. But before that, a large gathering of stakeholders, governments and UN bodies met in downtown New York at the New School University to discuss the Rio+20 priorities and enable a first set of discussions before the main process got going.

The conference was seen as a significant success, with over 200 participants and sessions on the green economy, sustainable development governance and scenarios for the future.

The prepcom itself turned out a disappointment. The bureau had decided not to have high-level panels to kick off the discussions but to go into discussions and statements from governments straight away.

The general attitudes of the governments and NGOs towards the issues started to become clearer.

The discussion on the green economy identified a number of areas that would need extra work:

- Was green growth interchangeable with green economy? At this point there was no definition of either.

- A Green Jobs strategy provided potentially better job creation opportunities with much more labour-intensive patterns than brown growth. However, any transition would need to be a 'just transition'.
- There was no 'one fits all' way of building the green economy. It would benefit a range of constituencies, and have a variety of manifestations, dependent on national circumstances.
- There should be an ecosystem valuation in the future that emphasises the full costs of ecosystem destruction.
- There would need to be a trade-supporting international environment without protectionism and conditionality.
- There will need to be some principles to guide transition to a green economy, roadmap and tool-kit.
- The present stimulus packages should help to inform what finance can do to help shift to a green economy.
- On sustainable development governance there was a call for strengthening the Commission on Sustainable Development and making it more action orientated, and it should work more at integrating its work within the UN system and with the international financial institutions (IFIs).
- There should be interagency coordination – greater UN cooperation, and an interagency coordinating mechanism, a 'Unified framework on sustainable development'.
- International environmental governance that should address the incoherent duplication of reporting burden.
- There was support for National Councils on Sustainable Development as a national way to promote integration.

MDG Review

The ten years of activities on the MDGs had achieved some success, according to the UN's Millennium Development Goals Report 2010. In the report's foreword, Secretary-General Ban Ki-moon said:

> This report shows that the Goals are achievable when nationally owned development strategies and policies are supported by international development partners. At the same time, it is clear that improvements in the lives of the poor have been unacceptably slow, and some hard-won gains are being eroded by the climate, food and economic crises. Billions of people are looking to the international community to realize the great vision embodied in the Millennium Development Goals. Let us keep that promise.
>
> (United Nations, 2010)

There had been big gains such as for children going to primary schools in many poor countries and strong progress in the areas of AIDS, malaria and child health. There was also some hope that the water goal would also be met. The share of developing countries' people who subsist on less than US$1.25 a day had fallen by 2005 on the base year of 1990 from 46 per cent to 27 per cent and was expected to drop to 15 per cent by 2015.

Other goals were off-target, and now having to be factored in was the impact of climate change and the international financial crisis. Many of the gains stood a serious chance of falling back in the following years as climate-change impacts on water availability and food production started to take hold. Added to this, the increased population expected in the coming years would increase pressure on resources and unless serious action was taken the future looked bleak.

The summit did generate money to address the MDGs in the area of reducing child mortality and improving maternity health. Around US$40 billion was pledged. The Asian Development Bank made a commitment of US$6 billion over three years on clean energy. The World Bank increased its support for agriculture by between US$6–8 billion over three years. This could be linked to the wakeup call that the CSD had given in 2009 to the low level of investment in agriculture which had since gained some momentum to be addressed. Japan also committed US$3.5 billion this time for education in developing countries, and the World Bank gave zero interest grants for basic education of up to US$750 (UN DPI 2010).

Even in times of financial crisis, governments were able to find funds for the MDGs. The question in 2015 will be whether they spent the money as things got a lot worse than they expected.

Nagoya

Perhaps because of the UNFCCC failure the previous December, governments needed a success; and without extensive pressure and media attention, Nagoya's meeting of the Convention on Biological Diversity (CBD) delivered that. It was still attended by over 18,000 representatives of governments, the UN, stakeholders and the press.

The meeting had achieved three interlinked goals. These were the adoption of a new ten-year Strategic Plan which would guide international and national efforts to save biodiversity, a resource-mobilization strategy to provide the way forward to a substantial increase to current levels of official development assistance in support of biodiversity; and the Nagoya Protocol on access to and sharing of the benefits from the use of the genetic resources of the planet.

Among the targets, the parties:

- agreed to at least halve and where feasible bring close to zero the rate of loss of natural habitats including forests;
- established a target of 17 per cent of terrestrial and inland water areas and 10 per cent of marine and coastal areas;
- through conservation and restoration, governments would restore at least 15 per cent of degraded areas; and
- would make special efforts to reduce the pressures faced by coral reefs.

(CBD, 2010)

Ahmed Djoghlaf, Executive Secretary of the Convention, said:

> History will recall that it was here in Nagoya that a new era of living in harmony
> was born and new global alliance to protect life on earth was established. History
> will also recall that this would not have been possible without the outstanding
> leadership and commitment of the government and people of Japan. If Kyoto
> entered history as the city where the climate accord was born, Nagoya will be
> remembered as the city where the biodiversity accord was born.
>
> (Djoghlaf, 2010)

Climate COP 16 – Cancún

After the catastrophe of Copenhagen, governments coming to Cancún wanted to
rebuild a base to address climate change. Yvo de Boer had left as head of UNFCCC
secretariat and was replaced by former Costa Rican negotiator Christiana Figueres.
The elements of the Cancún Agreements included:

- industrialized and developing country targets are officially recognized under
  the multilateral process;
- a total of US$30 billion in fast start finance from industrialized countries up
  to 2012;
- a new 'Green Climate Fund' to be set up under the Conference of the Parties;
- the establishment of a technology mechanism to increase technology
  cooperation and to support action on adaptation and mitigation.

UNFCCC Executive Secretary Christiana Figueres said:

> Cancún has done its job. The beacon of hope has been reignited and faith in the
> multilateral climate change process to deliver results has been restored. Nations
> have shown they can work together under a common roof, to reach consensus
> on a common cause. They have shown that consensus in a transparent and
> inclusive process can create opportunity for all.
>
> This is not the end, but it is a new beginning. It is not what is ultimately required
> but it is the essential foundation on which to build greater, collective ambition.
>
> (Figueres, 2010)

2010 ended with a number of multilateral successes – some big and some smaller.
In November, under the leadership of Executive Secretary Elizabeth Maruma
Mrema, the Parties to the Convention on Migratory Species (CMS) met. The 13
countries in which wild tigers live had seen their numbers plummet from 100,000,
a hundred years ago to around 3,600 today. The Tiger Summit agreed to work to
double tiger numbers by 2022, in its statement called the St Petersburg Declaration.

So it turned out to be a good year for multilateralism with the MDG+10, the
CBD, the CMS and UNFCCC agreements.

UNCSD – First Intersessional Meeting and 2nd PrepCom

The two-day meeting in January 2011 had been set up with key panels of experts to help create a richer debate. The meeting seemed to move beyond the call of the previous prepcom for a definition of the green economy, hoping that perhaps a set of principles might evolve as a way of helping governments each address the green economy in their own relevant way.

It was very clear that much of what had been promised in 1992 and 2002 hadn't been delivered, and there was a reluctance on the part of developed countries to spend much time on why that was the case. There was some understanding that the green economy might lead to more jobs and with the recession this was very attractive, but what was missing was any real data to support this.

On the sustainable development governance theme, much of the discussion revolved around support for strengthening UNEP.

In the area of the CSD, governments looked to strengthen the CSD perhaps going back to the original mandate. The strengthening of ECOSOC was also discussed to serve as an effective forum for multilateral discussion of the sustainable development agenda; using the structure of its Annual Ministerial Review as a model was suggested.

There was also discussion on strengthening national councils of sustainable development and involving finance and development ministries in them.

The second preparatory meeting in March benefited from a number of informal meetings on the green economy and institutional framework for sustainable development.

The weekend before the second preparatory session, UNEP organized a full-day consultation for governments on IFSD. Stakeholder Forum, the conference coordinator, used it to launch 'Sustainable Development Governance 2012' – a set of expert papers collected by ANPED, IUCN and the Climate and Sustainability Platform.

The coalition made a presentation. These papers which had been peer reviewed by the wider network provided real substance for governments to look at on reform. They included:

- 'Sustainable development governance towards Rio+20: framing the debate' by Jan-Gustav Strandeneas, Stakeholder Forum;
- 'Global governance in the 21st century: rethinking the environmental pillar' by Maria Ivanova, University of Massachusetts;
- 'Greening the international financial institutions (IFIs): finance for the next decade's sustainable development' by Kirk Herbertson, World Resources Institute;
- 'Environmental institutions for the 21st century: an international environmental court' by the ICE Coalition;
- 'Enhancing science – policy links for global sustainability' by Frank Biermann, Institute for Environment Studies;

- 'Embarking on global governance: thoughts on the inclusion of local' by Konrad Otto-Zimmermann, ICLEI; and
- 'Moving from principles to rights' by David Banisar, Article 19, Lalanath de Silva and Carole Excell, The Access Initiative, World Resources Institute.

In addition to addressing the green economy principles discussion that was prevalent at the previous meetings on 2012, Stakeholder Forum, the Earth Charter Initiative and Bioregional published a set of green economy principles drawn from the Stockholm Principles (1972), the Rio Principles (1992), the Johannesburg Political Declaration (2002), and a series of NGO initiatives on principles such as the Earth Charter, the trade unions coalition's Just Transition, the work of the New Economics Foundation, and Bioregional/WWF's One Planet Living. The paper was well received by governments as it drew heavily from text they had negotiated over the previous 30 or more years.

The discussion on the green economy was much more positive than it had been previously, and seemed to indicate the beginning of a consensus that a green economy might bring more coherence and balance to all three pillars of sustainable development. There were still concerns that the green economy might be a new form of protectionism, but there was recognition that Rio+20 could develop a set of options to benefit all countries. The recent draft of the UNEP report Towards a Green Economy suggested that:

> The greening of economies is not generally a drag on growth, but a new engine of growth.
>
> (UNEP, 2011a)

The report looked at examples of the benefits of green growth from developing countries themselves, including China, Uganda and Rwanda. The understanding was that the more practical evidence there is, the more likely those governments will embrace the issue and develop policy options.

Added to the discussion – by the small island states, in particular – was the issue of the 'blue economy'. Forty per cent of fish stocks in our oceans are over-exploited compared with only 20 per cent at the time of Rio in 1992. Even more stark is that 60 per cent of coral reefs have been destroyed or are at risk of destruction.

The discussion around the sectors which would be addressed by Rio+20 was starting to gain some clarity – energy, agriculture, water, urbanization, ecosystems and blue economy were at the top of the list.

On the issue of international framework for sustainable development, most of the discussion again resolved around IEG, but more and more governments were hungry for ideas on IFSD. The paper by Jan Gustav Strandenaes collected together a series of potentially effective ideas which included:

- the CSD being transformed into a Council of the UN General Assembly;
- the re-establishment of the Inter-Agency Committee on Sustainable Development; and

- the setting up of a Sustainable Development Board within ECOSOC for the UN 'One Country Programmes', a recommendation of the High-Level Panel on System Wide Coherence from 2007.

The reform of the CSD was on everyone's mind, but evidence that such reform would have to go far was about to get a strong boost.

The last CSD?

When CSD 2011 started there was much expectation for a successful meeting. There had been a series of intersessional meetings to prepare the input on the SCP issues which was thought to probably be the most controversial.

The session opened with a challenge to the chair and bureau over the ability of stakeholders to participate in the meeting. The plan put forward by the bureau allowed civil society interventions on the first day and then not again until the second week on the Wednesday. After pressure from all Major Groups and consultation with governments, this decision was changed to enable participation in all sessions. It seemed to represent a significant victory for the Major Groups, but still showed the continued shrinking of their available space.

Perhaps it also gave an indication of what was to come. The CSD was to fail to agree text for the second time in its history, so within four years, two CSDs had failed.

At the centre of the disagreements were three issues. The first was the rights of people in occupied territories. There was no willingness to follow past agreements by going back to refer to previous texts, and it reflected a rising tension around the Palestinian issue.

This discussion was not helped by the imminent arrival in Washington of Prime Minister Benjamin Netanyahu of Israel, at the politically motivated invitation of the US House of Representatives' Speaker, John Boehner, Republican of Ohio. This had been planned before the group Fatah announced a unity accord with Hamas. The day before Netanyahu arrived in Washington, President Obama announced that peace should be based on the 1967 borders with mutually agreed swaps. This was not something that Netanyahu would accept. US negotiators were focused on the need to not inflame the situation.

The second issue was the transitioning to a cleaner and more resource-efficient economy; this was a compromise for replacing 'green economy' but in the end G77 felt the term was as undefined as the green economy.

The final issue was that of means of implementation (MOI). G77 and China wanted to return to the old system used for the CSD, dealing with MOI, and putting it under each theme and then having a section of its own. There had been a trade-off but that didn't find its way into the final chair's text.

In the end, the issue that stopped the negotiations at 7.30am on the morning after the CSD session was supposed to close was Saudi Arabia calling for a roll call. Only 24 countries of the 53-member commission were present and the quorum was 27.

All the work on the SCP, mining and chemicals and the other issues were now lost, and the CSD closed, perhaps for the last time.

Solo, Indonesia meeting

In July 2011, the Indonesian government hosted a large meeting on International Framework for Sustainable Development. With governments realizing they needed to build common perspectives on governance, over 90 attended. A number of the suggestions in the papers tabled by NGOs earlier in the year were discussed in light of the previous CSD outcome.

Those reflected in the chair's summary were:

- Sustainable Development Council of the UN GA;
- the establishment of a World Environment Organization;
- support for national sustainable development plans within the UN Delivering as One framework;
- possible Inter-governmental Panel on Sustainable Development to which other science panels would report to enhance coherence; and
- support for Sustainable Development Goals.

(Indonesia, 2011a)

Sustainable communities – responsive citizens

The United Nations Department of Public Information (UNDPI) had been holding an annual conference for NGOs since the creation of the UN. The 2011 conference, its 64th, was scheduled to focus on Rio+20 and the tenth Year of Voluntarism. The conference became a major event for Rio+20 because of its timing in September 2011. Over 1500 NGOs and many UN agencies and programmes, some at executive director level, came together and produced a chair's text which identified some key outcomes:

- seeing Rio+20 and MDG+20 (2015) linked together;
- suggested seventeen Sustainable Development Goals for consideration;
- reform of UNEP and the UN CSD; and
- supported calls for conventions on the Rio Principle 10, Corporate Accountability and Responsibility and new technologies.

The meeting offered a real opportunity for discussion and a number of governments came to hear the ideas of NGOs.

Closing the conference, Felix Dodds, the chair, said:

Echoing the UN Secretary-General's call for a revolution, and Senator Robert Kennedy's call in a different time and on a different issue: A revolution is coming – a revolution which will be peaceful if we are wise enough; compassionate if we care enough; successful if we are fortunate enough – But a revolution is coming whether we will it or not. We can affect its character; we cannot alter its inevitability. This new world that we need to create will be based on sustainable societies and responsive citizens.

(Dodds, 2011)

Bogotá meeting on Sustainable Development Goals

Following up the Solo meeting, the governments of Colombia and Guatemala hosted a workshop in Bogotá to test the idea of Sustainable Development Goals (SDGs) as one of the main outcomes from Rio+20.

An outcome from the meeting was a paper outlining some of the benefits of the SDGs:

- build upon Agenda 21 and the Johannesburg WSSD Plan of Implementation;
- catalyse implementation at national and local levels, in response to national realities and priorities;
- assist the international community in focusing on key challenges that demand coherence and coordination at all levels;
- catalyse means of implementation at the international level and assist in identifying gaps and needs in developing countries;
- contribute to positioning the three pillars as cross-cutting building blocks for development throughout the UN system;
- complement the MDGs and enable stakeholders at all levels to work on concrete issues that focus on poverty eradication and on reducing environmental impacts from increased economic growth; and
- have universal application.

(Colombia, 2011)

They will play a significant role not just for Rio+20 but also for MDG+15.

The way forward

This book will go to press prior to Rio+20, but in late preparatory meetings it was clear that governments were struggling to bring some coherence to the discussions. Their chances of success – and possibilities of realizing the potential of the 2012 summit to to re-shape and re-mobilize support for the sustainable development agenda – remain very fluid. The following chapters will investigate some of the possible outcomes that those governments could be striving for during and after Rio+20 and the MDG+15.

Part II

The challenges of the future

Chapter 9

The governance gap

Changing course

Forty years ago the world's government began assembling a system of international conventions, agencies and programmes to address the demands of environment and sustainable development. For their first 30 years, those structures more or less effectively addressed the needs of general policy sectors or specific issue areas. Over the last 10 years, however, it has become frustratingly clear that the governance framework established so far simply is not sufficient.

The United Nations Environment Programme (UNEP), formed by the work of the Stockholm Human Environment Conference in 1972, and the United Nations Commission on Sustainable Development (CSD), formed by the Earth Summit in 1992, each achieved significant successes in their early days. The world, however, has changed drastically over the last 40 years, and both organizations – for different reasons and in different ways – seem no longer capable of delivering what they had been expected to or what is now needed.

International environmental governance

A year after UNEP was set up by the Stockholm Conference (1972) as the first international body on the environment, the European Union set up its Directorate General on the Environment, colloquially DG XI, the eleventh ministerial-level agency in the European Commission executive body (1973). Both UNEP and DG XI followed the model of the previously established US Environmental Protection Agency (EPA 1970) set up by Republican President Richard Nixon.

In its early years UNEP was at the centre of creating a wave of global and regional environment agreements to deal with the lack of regulation on a panoply of pressing environmental, natural resources and ecosystem issues.

As new conventions were agreed, secretariats for each were often set up in the countries that had hosted the convention meeting, each with its own governing body – and often structured to function at arm's length from UNEP.

By the time of the 'Rio+5' Special Session of the UN General Assembly (UNGASS) in 1997, there was a growing realization that the environmental agenda had been too fragmented.

At UNGASS, Brazil, Germany, Singapore and South Africa submitted a joint proposal to establish a world environmental organization. Addressing the GA, Germany's Chancellor Kohl described the structural and political imperatives:

> Global environmental protection and sustainable development need a clearly-audible voice at the United Nations. Therefore, in the short-term ... it is important that cooperation among the various environmental organizations be significantly improved. In the medium-term this should lead to the creation of a global umbrella organization for environmental issues, with the United Nations Environment Programme as a major pillar.
>
> (Kohl, 1997)

The attempts during that meeting were not successful, but governments continued to discuss how to consolidate and strengthen environmental coordination and provide a stronger voice to advocate for the environment.

By the opening of the World Summit on Sustainable Development (WSSD) in 2002, France had joined the call for a strengthened global environmental agency. In his speech to the Summit, President Jacques Chirac tied his support to the accomplishments of ten years earlier:

> To better manage the environment and ensure compliance with the Rio principles, we need a World Environmental Organization (WEO).
>
> (Chirac, 2002)

But the WSSD did not agree to a WEO, instead merely calling for the existing UN organizations to 'strengthen their contribution'. Reflecting continued concerns by G77 countries that any strengthened environmental enforcement capacity could be at the expense of their own development priorities – or could result in 'conditionalities' that restricted poorer countries' exports – the Johannesburg Plan of Implementation (JPoI) did not even mention 'environmental governance', but instead framed its call in terms of sustainable development priorities:

> The United Nations Environment Programme (UNEP), the United Nations Centre for Human Settlements (UNCHS), the United Nations Development Programme (UNDP) and the United Nations Conference on Trade and Development (UNCTAD), within their mandates, should strengthen their contribution to sustainable development programmes and the implementation of Agenda 21 at all levels, particularly in the area of promoting capacity-building.
>
> (United Nations, 20002c:, para 155)

In July and August 2003, nearly a year after the summit – an extraordinary heat wave swept across Europe for two weeks, resulting in an unprecedented 35,000 deaths. Almost 15,000 of those were in France alone, as elderly people in particular struggled against temperatures as high as 40°C (104°F) in a region that had never required the installation of air conditioners, in buildings built to retain, not disperse, the day's heat. Germany lost 7,000 to the heat wave, while Spain and Italy 4,000 each. London recorded its first 100°F temperature ever.

Five weeks later, President Chirac appeared in New York at the UN General Assembly high-level session, and framed the broader reason that the French government intended to take forward the idea of a United Nations Environment Organization (UNEO):

> Against the chaos of a world shaken by ecological disaster, let us call for a sharing of responsibility, around a United Nations Environment Organization.
>
> (Chirac, 2003)

Again unsuccessful in gaining enough votes to pass a resolution, France continued working to build a coalition, this time creating a 'Friends of UNEO' group which began to play a role in preparations for the World Summit on the MDGs and other emerging issues, in 2005.

The group observed that:

- there are problems of coherence and efficiency because of the increasing number of multilateral environmental agencies, resulting in a fragmentation of efforts;
- there are significant gaps in scientific expertise, early warning systems and information provision;
- the specific needs of developing countries are not adequately taken into account;
- financing for the environment at the international level is marked by instability.

In March 2005, UN Secretary-General Kofi Annan joined the call, recommending the establishment of an integrated structure for environmental standard setting to be built on existing institutions such as UNEP and the treaty bodies (Annan, 2005).

The General Assembly World Summit that September renewed interest in strengthening international environmental governance. The summit agreement started a new process:

> 169. Environmental activities:
> Recognizing the need for more efficient environmental activities in the United Nations system, with enhanced coordination, improved policy advice and guidance, strengthened scientific knowledge, assessment and cooperation, better treaty compliance, while respecting the legal autonomy of the treaties, and better integration of environmental activities in the broader sustainable development framework at the operational level, including through capacity-building, we agree to explore the possibility of a more coherent institutional framework to address this need, including a more integrated structure, building on existing institutions and internationally agreed instruments, as well as the treaty bodies and the specialized agencies.
>
> (WSSD, para 169)

It resulted in an informal General Assembly process begun in 2006, co-chaired by the Mexican and Swiss Ambassadors. In June 2007 they produced an options paper that called for the continuation of the discussion on whether to upgrade UNEP to a United Nations specialized agency. Their text, 'Informal Consultative Process on the Institutional Framework for the United Nations' Environmental Activities', otherwise known as the 'Building Blocks' paper, also included a number of incremental reform suggestions that were unilaterally adopted by UNEP, such as the creation of the position of Chief Scientific Officer.

Stakeholders responded to the 'Building Blocks' paper with suggestions of their own, in a paper titled, 'Options for Strengthening the Environmental Pillar of Sustainable Development – a Compilation of Civil Society Proposals on the Institutional Framework for the United Nations Environmental Activities':

> A number of principles should be adopted for a strengthened UNEP, namely: broad societal consensus on a long term vision for UNEP; reliable analysis of the present situation and future scenarios for UNEP; integrated planning comprising all dimensions of sustainable development; building on existing strategies and processes; increasing links between national and local level strategies; integration into financial and budget planning; early monitoring to steer processes and track progress; and effective participation mechanisms. In strengthening UNEP consideration must be given to the specific needs of developing countries and respect of the fundamental principle of 'common but differentiated responsibilities'. Developed countries should promote technology transfer, new and additional financial resources, and capacity building for meaningful participation of developing countries in IEG. Strengthening of IEG should also occur in the context of sustainable development and should involve civil society as important stakeholder and agent of transformation.
>
> (Stakeholder Forum et al., 2007a)

The government co-chairs then produced a draft resolution for consideration by all governments, but felt they did not have enough support to table the resolution for debate. Instead, they asked UNEP to take up the process they had started.

For several years, momentum on the process then waned.

At its Governing Council in February 2009 UNEP held a plenary session on international environmental governance. South African Minister Marthinus van Schalkwyk outlined a roadmap on IEG:

> The first milestone will be when we meet in a year from now, in February 2010. At that meeting we should ideally adopt a Ministerial Declaration on the principles and objectives (on IEG) that will guide our further work in the run-up to Rio plus 20.
>
> (van Schalkwyk, 2009)

The UNEP Governing Council set up an ad hoc ministerial process which became known as the Belgrade process. At its November 2009 meeting in Helsinki, ministers adopted the 'six functional objectives framework' (in a probably unintentional parody of Confucian teaching). The objectives were:

1 strengthen the science–policy interface;
2 develop a UN system-wide strategy for the environment;
3 realize synergies between multilateral environmental agreements;
4 link global environmental policy making and financing;
5 develop a system-wide capacity-building framework for the environment;
6 strengthen strategic engagement at the regional level.

Meanwhile, during the Belgrade Process-options for broader institutional reform were put forward, including:

1 enhancing UNEP;
2 establishing a new umbrella organization for sustainable development;
3 establishing a specialized agency such as a world environment organization;
4 possible reform of ECOSOC and the Commission on Sustainable Development;
5 enhancing institutional reforms and streamlining present structures.

Then, at the second meeting of the UNEP ministerial process, in November 2010 in Helsinki, the functional objectives were reduced to five:

1 creating a strong, credible and accessible science base and policy interface;
2 developing a global authoritative and responsive voice for environmental sustainability;
3 achieving effectiveness, efficiency and coherence within the United Nations system;
4 securing sufficient, predictable and coherent funding;
5 ensuring a responsive and cohesive approach to meeting country needs.

Governments backing reform were increasingly convinced that re-structuring of IEG was a critical necessity. In a paper for Helsinki, the co-chairs had recommended that UNEP focus only on:

1 enhancing UNEP;
2 establishing a specialized agency such as a world environment organization;
3 enhancing institutional reforms and streamlining present structures.

This was endorsed by the UNEP Governing Council in February 2011.

By that time, the issue of governance had officially been taken up by the Rio+20 process under its primary agenda item Institutional Framework for Sustainable Development. The discussion on IEG was intensifying.

An informal process on follow up to the UNEP 2011 Governing Council, under the co-chairs of Spain and Uruguay, met in June 2011. Its chairs' summary broadened the discussion to the economic and social pillars, regional and country level:

- It is evident that Member States are now strongly interested in exploring IFSD more broadly beyond IEG, and in an integrated manner. A takeaway message is that there is no need for separate future discussions on IEG.
- There is a gap in terms of who would facilitate future informal discussions on IFSD. This presents an opportunity for the Secretariat to raise this issue with the Bureau, and organize future dialogues/brainstorms.
- Separating IFSD issues would help open up for more diversely composed panels. The Secretariat could organize a series of seven informal dialogues/ brainstorms in New York: one for each of the three remaining Nairobi– Helsinki options; one on the economic pillar; one on the social pillar; one on the regional level; one on the country level.

(UNCSD Secretariat, 2011)

Incremental change

Against this background a number of options are now on the table; even without a complete re-structuring, there are options for IEG reform that could take place within existing structures which would also provide benefits beyond the governance issue. Maria Ivanova suggested that UNEP might be strengthened within its present mandate by:

- assigning UNEP the explicit task to serve as a global environmental information clearinghouse;
- focusing UNEP's capacity building programme;
- strengthening and utilizing the Environmental Management Group;
- coordinating developments increased coherence in IEG; and
- consolidating financial accounting and reporting.

(Ivanova, 2011)

These could all be accomplished as a way of incremental change within UNEP, making it more effective within its present mandate.

A NEW INSTITUTION

Is incremental change sufficient to address the challenges that the environment is facing and could be facing in the immediate future? Tinkering with the environmental architecture does not enable the multilateral system to both respond to and be proactive in addressing present and future challenges. It also doesn't produce a common vision of the emerging critical challenges and what should be done to address them.

Is there a way to begin to build new multilateral structures even before the decision for wholesale reform is agreed? One strategy already underway is to start physically 'clustering' the environmental conventions around common themes. The purpose would be to enable them to address mutually relevant issues and achieve administrative efficiencies. The three conventions relating to chemicals (Basel, Rotterdam and Stockholm) have taken the lead. In 2008, the secretariats of conventions on hazardous waste, prior informed consent and persistent organic pollutants decided to meet in joint session to address common issues. The first 'super COP' on chemicals took place in Bali, parallel to the UNEP Global Environmental Ministers Forum, in February 2010. The outcome recommendations included recruiting a joint head of the three conventions and to establish joint financial and administrative support service,

The next most obvious grouping to explore could be the biodiversity conventions. Other likely categories could include conventions related to atmosphere, oceans and fisheries, and freshwater and land. This approach holds significant potential. It could be accelerated as part of negotiations on the institutional framework at Rio+20.

Programme or agency

The discussion over the last 15 years on the creation of a World Environment Organisation (WEO) has produced a range of suggestions on its substantive focus, its functional objectives (e.g. that it be a strongly science-based organization), and its structure.

The upgrading of UNEP to an agency has the support of the European Union and most African countries. The basic differences between a programme and a specialized agency are that an agency:

- elects its own head (as opposed to being appointed by the UN Secretary-General);
- receives assured dues based on the agreed UN scale of percentages paid by all countries and not just the voluntary amounts decided each year by those countries that choose to participate at its meetings – and therefore significant financial stability.

One of the critical – if not universally supported – rationales for creating a stronger, more coherent or more autonomous environmental organization is to provide it with the institutional tools to balance the World Trade Organization (WTO). The WTO is the great exception in the world of multilateral institutions – a body that has accrued immense power as a negotiating, coordinating and judicial authority. Given the increasing number of areas where environmental and economic areas overlap and the tendency so far for WTO rulings to champion economic interests at the expense of environmental, there is a powerful argument for establishing another organization that can represent environmental priorities and interact with the WTO from a position that at least approaches political parity.

One suggestion on the structure of a WEO is to establish a committee system similar to that of the WTO. That could be accomplished through clustering the present conventions into thematic areas, such as:

- chemicals;
- biodiversity;
- atmosphere;
- oceans and fisheries; and
- freshwater- and land-related conventions.

A study by the German Advisory Council (2001) contended that the Conferences of the Parties (COPs) of the various MEAs can be brought under the umbrella of a WEO to create bodies that are coordinated but that function with a 'high degree of autonomy', in the same way as special committees of the WTO Ministerial Conference.

Others argue that such an analogy is inapt, because almost all of the WTO committees are committees of the whole, and none of them so far has operated with any autonomy from the WTO membership as a whole. The only regime that has consolidated in the way those proponents would like for a WEO is the World Intellectual Property Organization (WIPO) (Charnovitz, 2002).

That, of course, would still provide a precedent.

Ideally, some say, the WEO should be given duties that distinguish it from the national environmental agencies that exist in each country. Otherwise, the world agency would duplicate the national agencies (Charnovitz, 2002).

But all existing international agencies overlay the agendas of national agencies. No existing major international agency looks only at global problems. The mandate of the WTO, the ILO, the WHO, the FAO and others also allows them to deal with issues at the country level.

One prominent academic, Frank Biermann, suggests three possible models for a WEO:

Model 1: Upgrading UNEP to a specialized agency with organizational status. The agency would facilitate norm-building and norm-implementation processes and be given additional legal and political powers, which would allow it to adopt drafts of legally binding treaties negotiated by sub-committees.

Model 2: Integrating several existing agencies and programs into one world environment organization. One of the ideas suggested is to have the organization follow the model of the World Trade Organization, which has integrated diverse multilateral trade agreement.

Model 3: Creating a hierarchical intergovernmental organization on environmental issues that has majority decision-making and enforcement powers. One common cited example is the Hague Declaration of 1989, which was moving towards becoming an international agency with sanctioning powers.

(Biermann, 2011)

He underlines the reasons for a WEO/UNEO:

the current system of international organization and of international bureaucracies lacks effectiveness. This is partially due to a lack of standing of the core agencies in this respect and the overall fragmentation of earth system governance. The establishment of a UN Environment Organization would improve coordination of earth system governance; pave the way for the elevation of environmental policies on the agenda of governments, international organizations and private organizations; assist in developing the capacities for environmental policy in African, Asian and Latin American countries; and strengthen the institutional environment for the negotiation of new conventions and action programs as well as for the implementation and coordination of existing ones.

(Biermann, 2011)

One problem is that the US may not be able to join such a new agency. This would require 60 votes within the US Senate which could be impossible at the present time. As the US is the largest funder of UNEP this would also have significant impact. Finally governments will have to address the clustering of MEAs and how to proceed when the relevant MEAs are under the umbrella of other UN agencies such as IMO, ILO, UNESCO or FAO. A possible middle way might be for the UN General Assembly to agree to provide UNEP with universal membership and a UN scale of dues, without going so far as declaring it an agency – something which could be left to a future decision.

Some additional options might be considered by governments:

1 reviewing mandates of UN agencies to consider simplifying areas of common interest;
2 establishing a joint secretariat role between a WEO and another UN agency; and
3 changing the secretariat from one UN agency to another.

Global norm-building and institutionalization could be strengthened by following a model similar to that of the International Labour Organization. By following such a model the general assembly of the UNEO would have the ability to adopt draft treaties that have been negotiated by subcommittees and other legal instruments on its own. Such an agency would address the issue of individual legislative bodies exercising their strong centrifugal tendencies and overlooking their links with other fields. The overlap in the functional areas of institutions would diminish because the agency would influence the creation of a clear strategy to ensure worldwide environmental protection. It would provide a level of streamlining and harmonization that would reduce the current lack of coordination in the reporting system. Further, it would address the issue of developing countries being unable to attend meetings by facilitating the development of 'environmental embassies' which would reduce the costs incurred by countries and increase their negotiation influence.

International Court for the Environment

Many Multilateral Environmental Agreements (MEAs) are simply declaratory initiatives, rather than legally binding treaties such as Agenda 21, and agreements at the Commission on Sustainable Development. Such initiatives are agreed because states – or at least a sufficient blocking minority of states (which in a consensus negotiating process doesn't have to include many) – might want to adopt lofty aspirational statements of principle (or feel the political need to act like they do), but are sceptical about adhering to global agreements that may come into conflict with their perceived national interest. Or – in the case of certain political or economic super-powers – they might perceive it in their national interest to oppose *any* global agreement or instrument that is beyond their immediate national control.

That is not to say that declaratory initiatives don't hold significant value. They do. They can make it normative for nations to engage in – or prohibit – specific policy behaviour. They frequently serve as a model for national legislation or regulation by governments who agree with their goals. They can serve as a model for other governments who might not care about their goals but who seek financial assistance from those who do.

> The Environmental Protection Committee of the National People's Congress has recently been drafting two pieces of environmental law – a general framework law and a law on water pollution management. What is remarkable in the drafting of these new laws is that when the drafters from the National People's Congress could not agree about a particular rule or principle to include, the drafters would bring out the Chinese language version of Agenda 21 and lift the language directly from the Agenda and incorporate it into the draft text.
> (Sands, 1997)

This is a good example of how international agreements can be transformed into national legislation.

MEAs build a foundation for continuing negotiations that can eventually lead to legally binding treaties, with force-of-law status, and monitoring, compliance and reporting mechanisms.

The fact that the timeline for achieving such treaties is typically measured in years or decades, can be deeply frustrating for those directly impacted by the issue (or for their advocates), and is the source of much scepticism, and from some factions, ridicule, regarding the usefulness of multilateral processes.

But the fact they *can* eventually lead to effective legal agreements is precisely the reason that so many governmental and non-governmental representatives – who could be spending their energy and resources on far more immediate pursuits – instead decide to spend those necessary years or decades toiling in the at-times drought-stricken vineyards of intergovernmental negotiations.

A faster track – treaties without trauma

A few of the Multilateral Environmental Agreements have specific compliance mechanisms, but even these are relatively weak, as few countries are prepared to submit themselves to an independent review of compliance with the convention requirements. There has also been a tendency in the last 20 years for some environmental agreements to use trade mechanisms in order to ensure compliance (e.g. Montreal Protocol, Basel Convention on the Trans-boundary Movement of Hazardous Waste, Convention on International Trade in Endangered Species of Wild Fauna and Flora (CITES) and Prior Informed Consent (PICs). This perhaps provides an interesting argument for stronger enforcement mechanism for future conventions.

An International Court for the Environment (ICE)

But to be effective, any multilateral agreement needs to include mechanisms to ensure reporting, monitoring and compliance. The area of sustainable development currently features a mechanism for reporting at the CSD, but it is a weak mechanism that involves only voluntary reporting, with no provision even for verification.

There is currently no mechanism for compliance and enforcement in sustainable development, as exists in other areas, such as war crimes or trade.

There are a number of international courts, tribunals and arbitral bodies that have been created to decide on states' obligations and responsibilities under international environmental law. There are none in the current system that delivers sufficient access to justice for non-state actors or provide a forum that is suitable to hear very technical scientific evidence common to environmental cases.

The present deficit would be addressed by an International Court for the Environment (ICE):

> It is envisaged that the ICE would become the principal court dealing with international environmental law, helping to clarify existing treaties and other international environmental obligations for states and for all other parties including trans-national corporations. It would do this through dispute resolution, advisory opinions, and the adjudication of contentious issues presently unclear or unresolved.
>
> (ICE Coalition, 2011)

There are grounds under Principles 21 and 22 of the Stockholm Declaration dealing with trans-boundary pollution where it would be easy to imagine a class legal challenge through domestic courts in the polluting countries for greenhouse gasses.

An ICE would enable a more coherent approach to this. It would need to have the ability to recommend sanctions against states that refused to act in accordance to the Courts decision. This would be very controversial but would offer some fairness and equity in the world.

A government could in principle bring a grievance against another government for violations of environmental or sustainable development treaties to the International Court of Justice (ICJ), or colloquially, the World Court.

This may occur soon, as governments in the Alliance of Island States (AOSIS) are considering whether to bring legal action against those responsible for producing global warming.

Only states are allowed to bring cases before the ICJ. There is no provision for non-state actors like cities, or businesses or NGOs to bring cases against states and other non-state actors. The ICJ provides them no legal standing.

Additionally, the slow emergence, fragmentation and a lack of implementation of MEAs have caused much difficulty in the enforcement of international environmental law.

Such deficits in sustainable development governance are the motives for seeking the creation of an ICE. By creating such an organization, governments would assure there will be an enforcement of trust so that states are no longer contemptuous of following global agreements that aren't immediately convenient, or afraid of adhering to standards that might cost them economically if other states fail to comply.

With an environmental court, when MEAs become legally binding treaties there will be a mechanism for compliance and enforcement, which would ensure that states compete on a 'level playing field'. In other words, it will ensure that countries adopt and implement national environmental protection standards that measure up to the global standard – or that other countries can hold them to it (ICE Coalition, 2011).

It will also create scientific expertise at the international judicial level – important because of the need to assemble a chamber of judges knowledgeable enough on matters of environmental science to participate in decisions that can significantly influence future state political and economic practice, and to impact on customary international law.

Finally, an ICE could pave a path for non-state actors to have access to an international judiciary in a forum where environmental scientific evidence is both respected and understood.

Scientific fragmentation

The establishment of the Intergovernmental Panel on Climate Change (IPCC) was a model-setting and extremely successful initiative for collecting and organizing vast amounts of scientific inputs to an environmental governance process that yielded an unimpeachable and highly usable series of results.

But with the establishment of the more recent Intergovernmental Panel of Biodiversity and Ecosystem Services (IPBES) and growing calls for scientific bodies similar to the IPCC and IPBES for chemicals, oceans and seas, and desertification, there is a danger that just as there was fragmentation of the MEAs, there may now develop fragmentation of the intergovernmental science bodies.

One option for avoiding that could be establishing an over-arching Intergovernmental Panel on Sustainable Development (IPSD) under which present panels and any future panels would sit. Such an IPSD should report to the follow-up mechanism agreed at UNCSD 2012.

An IPSD would enable the development of common data collection and the clear identification of gaps, but more vital to policy makers, the option to build models that integrate the present and future threats and paint a clear picture of what the results might look like if multiple impacts from unrelated environmental and development crises coincide to cause escalating and mutually-reinforcing real-world catastrophes.

The IPSD has been promoted by Indonesia within the Solo Message as the outcome from their IFCSD meeting in 2011.

Sustainable development governance

Established at Rio in 1992, the UN Commission on Sustainable Development has had a mixed history.

Its original mandate included:

> 3(a) To monitor progress in the implementation of Agenda 21 and activities related to the integration of environmental and developmental goals throughout the United Nations system through analysis and evaluation of reports from all relevant organs, organizations, programmes and institutions of the United Nations system dealing with various issues of environment and development, including those related to finance.
>
> (United Nations, 1992b)

The CSD achieved notable successes in its first eight years. Over those years it became a venue for open dialogue among governments on the full range of sustainable development issues. It allowed cross-sectoral consideration of the broader needs of policy and implementation. It established extraordinary and model-establishing access to non-state participants from multiple stakeholders.

The CSD provided a common venue for the Rio conventions and UN agencies and programmes to report to on progress. And it stimulated considerable energy around the follow up to Agenda 21, at all levels, serving in effect as an annual 'home' for the international sustainable development 'family' to meet at, strategize and exchange experiences during those early years.

But it increasingly revealed it lacked the tools – and the cooperation of governments and other intergovernmental bodies – to mobilize the translation of

---

Box 9.1 Successes of the UN Commission on Sustainable Development

1994 CSD called for the development of an 'effective legally binding instrument concerning the Prior Informed Consent (PIC) procedure on the importation of chemicals'.

1995 CSD established the United Nations Inter-government Panel on Forests.

1996 CSD set out the requirements for the establishment of the institutional arrangements for the implementation of the Global Programme of Action for the Protection of the Marine Environment from Land-based Activities.

1997 UNGASS called for, by the year 2002, the formulation and elaboration of national strategies for sustainable development, the establishment of the UN Intergovernmental Forum on Forests (IFF) for three years (1997–2000), and the establishment of multi-stakeholder dialogues with governments within the UN CSD.

1998 CSD called on UNCTAD, UNEP and UN DESA to help develop a vulnerability index for the quantitative and analytical work on the vulnerability of small island developing states, and the establishment of a review of voluntary initiatives within industry.

1999 CSD established an expansion of the United Nations guidelines on consumer protection to include sustainable consumption. It also established an open-ended informal consultation processes on oceans and seas under the UN General Assembly.

2000 CSD set out the terms of reference for a new permanent body – the United Nations Forum on Forests.

---

the sustainable development principles it was agreeing into actual implementation within individual countries, on the ground.

Even at the Five-Year Review in 1997, questions were being asked about the effectiveness of the CSD in being able to mainstream sustainable development including within the UN system.

As a functional commission of ECOSOC, the CSD had no power to require action from its decisions, particularly as they relate to giving instructions to UN agencies, programmes and funds. Unlike UN agencies, it did not report to the UN General Assembly and it did not have sufficient financial resources.

If we accept the fact that sustainable development is an overarching principle, the question becomes why the CSD was placed within the ECOSOC framework, as a functional commission. This seems very contradictory: if sustainable development is an overarching principle, then ... why is the CSD not given a 'higher' place in the hierarchy of the UN system? Logically speaking, the CSD should be in a position to give political guidance to and integrate the sustainable development policy and work not only of ECOSOC and it commissions, but of all the agencies, programmes and bodies of the UN system.

(INTGLIM, 1997)

The Johannesburg Summit (WSSD) in 2002 recognized that sustainable development needed to be strengthened within the UN system:

145. ... Although the role, functions and mandate of the Commission as set out in relevant parts of Agenda 21 and adopted in General Assembly resolution 47/191 continue to be relevant, the Commission needs to be strengthened, taking into account the role of relevant institutions and organizations. An enhanced role of the Commission should include reviewing and monitoring progress in the implementation of Agenda 21 and fostering coherence of implementation, initiatives and partnerships.

(United Nations, 2002b: para 145)

It also recognized that sustainable development governance had to be far wider than the CSD, and include the entire UN system:

143. The General Assembly of the United Nations should adopt sustainable development as a key element of the overarching framework for United Nations activities, particularly for achieving the internationally agreed development goals, including those contained in the Millennium Declaration, and should give overall political direction to the implementation of Agenda 21 and its review.

(United Nations, 2002b: para 143)

The most recent phase of the CSD (2003 to 2011) has been its most difficult, with two of its sessions (2007 and 2011) failing to agree any outcome at all. This has intensified calls for its reformation.

A Sustainable Development Council

The original discussion on sustainable development governance occurred during the Earth Summit in 1992. As mentioned previously, the Secretary-General of the Rio Conference sought the transformation of the UN Trusteeship Council – one of the six principal organs of the United Nations – into a broader Trusteeship Council for the global commons.

Especially relevant therefore to global stewardship would have been the transformation of the Trusteeship Council. Its original function was to act as trustee for former colonies in the process of becoming independent. As this role had now been almost entirely fulfilled, it would make sense to give the Trusteeship Council new life as the forum in which the nations of the world come together to exercise collectively their stewardship for the integrity of the environment, resource and life systems of our planet and the commons beyond national jurisdiction.

The transformation of the Trusteeship Council was supported after Rio by the Commission on Global Governance (CGG) in its 1995 report, *Our Global Neighborhood*:

> The Trusteeship Council should be given a new mandate over the global commons in the context of concern for the security of the planet.
>
> (United Nations, 1995)

Two years later, in 1997, then Secretary-General Kofi Annan referred to what might be termed 'a new concept of trusteeship' in an explicitly environmental context. In a report called *Renewing the United Nations: A Programme for Reform* the UN Secretary-General proposed that the Council

> be reconstituted as the forum through which Member States exercise their collective trusteeship for the integrity of the global environment and common areas such as the oceans, atmosphere and outer space. At the same time, it would serve to link the United Nations and civil society in addressing these areas of global concern, which require the active contribution of public, private and voluntary sectors.
>
> (UN Secretary-General, 1997)

Such reasoning would link collective trusteeship with emerging concepts of global governance (Strandenaes, 2011).

The primary reason why the Trusteeship Council idea wasn't taken up by governments was the argument by some that it would require a UN Charter amendment. The suggestion was also being made at the same time that suggestions were being considered for re-structuring the UN Security Council – and certain governments who were opposed to Security Council reform did not want to create a precedent.

Norway, under Prime Minister Gro Harlem Brundtland, also tried to seek a stronger body in 1992 when it put forward that any UN Commission on Sustainable Development should report to the UN General Assembly and not ECOSOC.

Meanwhile, the CSD was functioning and achieving successes. Much of the original suggestions for the role of re-structured Trusteeship Council – e.g. overseeing the global commons such as oceans, the atmosphere and space – seemed fulfilled by the CSD.

Of course, the previous two obstacles no longer apply. So while much of the attention of governments may have shifted, early advocates are reminding policy makers of the pragmatic advantages – and the powerful symbolic elegance – of a Trusteeship Council that protects the entire planet and its species.

Most recently, a number of NGOs have begun advocating the creation of a Sustainable Development Council of the UN General Assembly (similar to the 1992 idea from Norway).

The strength of this idea is that it inarguably would not require a Charter amendment and has a precedent in the recent establishment of the Human Rights Council set up out of the World Summit 2005. Its text could provide a model for the formation of a Sustainable Development Council:

> 57. Pursuant to our commitment to further strengthen the United Nations human rights machinery, we resolve to create a Human Rights Council.

> 159. The Council should address situations of violations of human rights, including gross and systematic violations, and make recommendations thereon. It should also promote effective coordination and the mainstreaming of human rights within the United Nations system.

> 160. We request the President of the General Assembly to conduct open, transparent and inclusive negotiations, to be completed as soon as possible during the sixtieth session, with the aim of establishing the mandate, modalities, functions, size, composition, membership, working methods and procedures of the Council.
> (United Nations, 2005)

One of the key outcomes from Rio+20 could be the adoption of a similar section of text. The critical question is what a Council of the UN General Assembly should address.

The idea of a Council was raised in a paper produced by UN DESA (2011) for the August 2011 High Level Dialogue on Institutional Framework for Sustainable Development (IFSD), in Indonesia. Attended by over 70 governments and many IGOs, its premise was then reflected in the 'Solo Message':

> At the international level, we need an organization to enhance the integration of sustainable development. Various options were discussed, ranging from an enhanced mandate for ECOSOC and reviewing the role of CSD, to the establishment of a Sustainable Development Council.
> (Indonesia, 2011b)

This might be a start, but the more critical agenda that needs a home is that of emerging issues and any sustainable development goals agreed at Rio. Since the Johannesburg Summit, issues such as climate, energy, food and water security, ecosystem and bio-diversity destruction, and massive natural disasters have found increasing space on our televisions and in our national political debates.

A Sustainable Development Council could provide for policy formulation around this environmental security agenda and the inter-linkages between the different issue areas. It could also include the emerging green economy, as well as issues that currently have only a weak institutional home, such as migration, natural disasters and urbanization. A Council could strengthen the UN system's ability to respond to this emerging agenda.

To establish such a Council, one long-time observer suggested that the Rio+20 Conference recommend:

- establishing a high level committee of experts to further develop the structure, mandate, and work-programme of the council, and provide it with the necessary authority within the UN system to allow the panel to consult with authority all levels of the intergovernmental system;
- making sure this high level committee of experts consists of representatives from all relevant stakeholders, representing governments, the intergovernmental system, the major groups and academia.

(Strandenaes, 2011)

And he recommended that the following elements be considered:

- the relationship of the new permanent council to the UN General Assembly;
- the structure of the council, leadership, secretariat, membership, meeting frequency, etc;
- the relationship to the other permanent councils, in particular ECOSOC;
- the relationship to other relevant UN entities, such as specialized agencies, subsidiary bodies working on sustainable development and other relevant intergovernmental institutions, in order to operationalize the three pillars of sustainable development: the economic, social, and environmental. A particular focus should be given to the financial institutions with a view to green economy and to UNEP;
- elaborate on the council's position to sustainable development and environmental governance as well as its relationship to environmental law systems, and the intergovernmental entities working on this;
- how emerging issues will find a proper place on the agenda of the new council, being aware of the fact that in some cases these issues will be of an unpredictable nature;
- how the open and interactive nature of the present commission on sustainable development can be adopted by the new council to allow for rich exchanges of ideas, where the major groups as envisaged by Agenda 21 is given an integrated role.

(Strandenaes, 2011)

An agreement by the Rio+20 Conference and recommendation to the General Assembly would provide a foundation for the rapid establishment of a functioning, effective UN Sustainable Development Council.

Reform of United Nations Economic and Social Council

If the new Sustainable Development Council of the UN General Assembly were to focus on emerging issues and SDGs then where should the follow up for Rio+20, JPoI and Agenda 21 take place ? And what form should it take?

There were a number of interesting experiments within the UN CSD in its first ten years which could be brought together into a new process within the UN Economic and Social Council. These would include:

- national reporting;
- involvement of national multi-stakeholder platforms;
- partnership learning processes;
- good practice replication.

The creation of a non-policy forum to review implementation at the ECOSOC level could have some interesting features. It could be:

- a problem solving forum;
- a forum with a governance model which includes stakeholders;
- a scaling-up of good partnerships;
- a peer group review of national reports (this was undertaken in the first five years of the UN CSD and is a feature of the ECOSOC Annual Ministerial Review of MDGs;
- an annual session of the national multi-stakeholder platforms.

By being a non-policy forum it could develop real momentum around implementation and addressing the obstacles to them. It could bring in relevant stakeholders such as parliamentarians to look at legislative obstacles and develop guidelines to help governments overcome them. It could work local authorities on addressing local planning regulations or financial frameworks to generate funding for non-profitable but essential risk management.

An ECOSOC forum for sustainable development could draw on the extensive range of stakeholder experience, and integrate it into implementation by governments.

UN interagency reform

The challenge at the global level of coherence and coordination is made more difficult by the nature of the institutions that are being coordinated, which have different governing bodies.

Agenda 21 in 1992 recommended the establishment of an Inter-Agency Committee on Sustainable Development (IACSD). Its section on high-level inter-agency coordination mechanism said:

38.16. Agenda 21, as the basis for action by the international community to integrate environment and development, should provide the principal framework for coordination of relevant activities within the United Nations system. To ensure effective monitoring, coordination and supervision of the involvement of the United Nations system in the follow-up to the Conference, there is a need for a coordination mechanism under the direct leadership of the Secretary-General.

38.17. This task should be given to the Administrative Committee on Coordination (ACC), headed by the Secretary-General. ACC would thus provide a vital link and interface between the multilateral financial institutions and other United Nations bodies at the highest administrative level. The Secretary-General should continue to revitalize the functioning of the Committee. All heads of agencies and institutions of the United Nations system shall be expected to cooperate with the Secretary-General fully in order to make ACC work effectively in fulfilling its crucial role and ensure successful implementation of Agenda 21. ACC should consider establishing a special task force, subcommittee or sustainable development board, taking into account the experience of the Designated Officials for Environmental Matters (DOEM) and the Committee of International Development Institutions on Environment (CIDIE), as well as the respective roles of UNEP and UNDP. Its report should be submitted to the relevant intergovernmental bodies.

(United Nations, 1992a)

The ACC established the Inter-Agency Committee on Sustainable Development (IACSD) in 1993, which was chaired by the Under Secretary-General for Policy Coordination and Sustainable Development (later to become UN DESA). The IACSD was made up of what were called Task Managers of Agenda 21. These were the agencies and programmes who were assigned responsibility for each of the different chapters of Agenda 21.

From 1993 to 2000 the IACSD dealt with coordination of the implementation of Agenda 21 throughout the UN system. The IACSD was abolished in 2000 under the UN reform package that transformed the ACC into the Chief Executive Board in 2001.

Since then there has been no interagency coordination on sustainable development. Instead UN Oceans, UN Water and UN Energy were created as interagency mechanisms to facilitate coordination and coherence in these three policy areas.

The Environment Management Group (EMG) was set up by the United Nations General Assembly in paragraph 5 of its resolution 53/242 in 1999.

The EMG consists of all the agencies in the United Nations system, and the secretariats of the multilateral environmental agreements, the World Bank, the International Monitory Fund (IMF) and the World Trade Organization (WTO). It was mandated to coordinate approaches and information exchange, promote joint action by United Nations agencies and create synergies among and between the activities of the UN agencies on environment and human settlement issues. UNEP serves as the secretariat for the group.

UNCSD 2012 could consider the re-establishment of an interagency mechanism for sustainable development, and linking it to the CEB, the three inter-agency policy areas and the EMG.

This could be done through a subcommittee of the Chief Executive Board. At present the CEB has three subcommittees:

- High-Level Committee on Programmes (HLCP);
- High-Level Committee on Management;
- United Nations Development Group.

Now is the time to establish a fourth subcommittee, a High-Level Committee on Sustainable Development. Such a mechanism might include its own subcommittees:

- UN Water
- UN Oceans
- UN Energy
- Environmental Management Group.

This would be a comprehensive way to ensure coherence and effectiveness of UN system.

Sustainable Development Board

The Secretary-General's High-Level Panel on System Wide Coherence (United Nations, 2006) recognized that a new governance mechanism was needed for operational coherence and the implementation of policies in the area of sustainable development at the country level. It recommended the establishment of a Sustainable Development Board to oversee the One United Nations country programmes.

It suggested that the role and mandate of the Board should mainly be to:

- endorse 'one country programmes' and approve related allocations;
- maintain a strategic overview of the system;
- review the implementation of global analytical and normative work of the UN in sustainable development activities;
- oversee the management of the funding mechanism for the MDGs;
- review the performance of the resident coordinator system;
- commission periodic strategic reviews of 'one country programmes';
- consider and act on independent evaluation and audit findings.

However most of this was not implemented. According to Mohamed El-Ashry:

> It has been four and a half years since the High-level Panel released its report, 'Delivering as One.' While some of the Panel's recommendations in the areas of 'one country program,' gender equality and business practice have been implemented, none of the recommendations related to environment and sustainable development have been implemented. It is simply a case of lacking political will.
>
> (El-Ashry, 2011)

Among the most significant recommendations of the Panel were that those meetings of the UN Sustainable Development Board should: 'Supersede the joint meeting of the boards of UNDP/UNFPA/gender entity, WFP and UNICEF'.

The Sustainable Development Board would report to a new 'sustainable development' section of the Economic and Social Council.

The original recommendation did not include UNEP, as its governing body is different in structure from the other entities. Should UNCSD outcomes recommend for UNEP to have an Executive Board, then it could be included within this Sustainable Development Board.

Such a Board could also play a significant role in coordinating and ensuring coherence in the implementation of UNCSD outcomes as well as Agenda 21 and the JPoI.

National coordination

More effective government coordination at the national level would have a great impact on the ability of the UN and IFIs to coordinate sustainable development policy at a global level.

In many countries, ministries of the environment are not among the most influential ministries of state. In most countries, it is the ministries of finance – which are often not the champions of sustainable development – which direct government policies in areas impacting on sustainable development, often without input from colleagues in more relevant ministries. This balance of control also pertains when national delegations are preparing their positions prior to attending IFI board meetings.

It might be useful to suggest that governments consider establishing a unit within the office of the prime minister or foreign minister to seek policy and strategic coherence among government departments operating within intergovernmental arenas.

Principle 10 of the Rio Declaration

The UN Economic Council for Europe (UNECE) followed up the Rio Conference by negotiating a regional convention on the Rio Declaration's Principle 10 . 'The

Convention on Access to Information, Public Participation in Decision-Making and Access to Justice in Environmental Matters', commonly known as the Aarhus Convention, was signed 25 June 1998 in the Danish city of Aarhus. It entered into force on 30 October 2001. It has been ratified by the European Union and 41 countries.

The Convention has three pillars:

1 Access to information: any citizen should have the right to get a wide and easy access to environmental information. Public authorities must provide all the information required and collect and disseminate them and in a timely and transparent manner. They can refuse to do it just under particular situations (such as national defence);
2 Public participation in decision making: the public must be informed over all the relevant projects and it has to have the chance to participate during the decision-making and legislative process. Decision makers can take advantage from people's knowledge and expertise; this contribution is a strong opportunity to improve the quality of the environmental decisions, outcomes and to guarantee procedural legitimacy;
3 Access to justice: the public has the right to judicial or administrative recourse procedures in case a Party violates or fails to adhere to environmental law and the convention's principles. It provides a real-world example of how multilateral principles can be developed further into regional and then domestic law. In contrast to the Rio Declaration, which is a soft law agreement, the Aarhus convention is a 'hard law' agreement that requires compliance.

Other regions of the world are now looking at the possibility of developing their own regional conventions on Principle 10. The furthest along in the process is the Economic Council for Latin America and the Caribbean (ECLAC). Most of the countries in the region have access to information enshrined in their constitutions (Costa Rica, Ecuador, Mexico and Peru), others through legislation (Brazil, Argentina and Ecuador).

In a paper prepared for the UNCSD Regional Preparatory Meeting for Latin America and the Caribbean, ECLAC recommended:

The proposal in this area is to establish a regional legal instrument guaranteeing the rights enshrined in principle 10 of the Rio Declaration. This could be based on lessons learned from the European experience regarding the Convention on Access to Information, Public Participation in Decision-Making and Access to Justice in Environmental Matters (Aarhus Convention).

Such an instrument should facilitate the creation of clear legal frameworks with procedures for obtaining the environmental information held by State bodies (including private sector utility companies). It should have oversight mechanisms and procedures for ensuring that vulnerable groups and indigenous peoples have access to information. The framework should include pollutant release and transfer registers (PRTRs) on the manufacturing industry.

(ECLAC, 2011)

If ECLAC are able to launch the process around Rio+20 then other UN Economic Commissions may follow suit and a global framework convention might be possible on Principle 10 by 2022, built from the bottom up.

An Ombudsperson for Future Generations

Taken as a whole, the current generation's way of life is already living so far beyond the earth's means of support – and still accelerating – that it is hurtling humanity towards a highly unstable and unequal world for future generations that will lead to disaster. With election cycles as short as two years in some countries, businesses reporting quarterly and a 24-hour news cycle, few are given the time to take into consideration the future needs of their own children and grandchildren, much less the needs of other species.

The culture we have is built on instant gratification – 'my one minute of fame'. Somehow we need to find a way to factor in the needs of the future and of others. One idea for attempting this institutionally is establishing a position of 'Ombudsperson for Future Generations'. Some countries are already trying this out. New Zealand has a 'Parliamentary Commissioner for the Environment', Israel a 'Parliamentary Commission for Future Generations', and Hungary also a 'Parliamentary Commissioner for Future Generations'.

> The Parliamentary Commissioner for Future Generations shall follow with attention, estimate and control the emergence of the provisions of the law ensuring the sustainability and improvement of the situation of environment and nature (hereinafter together 'environment'). It shall be his duty to investigate or to have investigated any improprieties he has become aware of relating to these, and to initiate general or particular measures for the redress thereof.
>
> (Hungary, 1993: 27/B)

Having national ombudspersons for the future will make someone responsible to evaluate policies and laws for the long term to see what impact they may have. It will help create a narrative which perhaps can transcend party politics.

> Changing the way we speak about the challenges we face is helping citizens re-engage with policy making. Choices become more important when it is clearly communicated how the lives of future generations – people's own children and grandchildren – are affected. Such a change of perspective reconciles the current generation's hopes and desires with those of the generations to come and gives SD its original meaning back.
>
> (Göpel, 2011)

Box 9.2 **Criteria for effective representation of future generations**

- Independence: The ombudsperson and his or her staff should not hold another governmental post that would influence their freedom or reasoning. Ideally, the office should be legally independent and its budget predictable over longer time spans.
- Legitimacy: The office should enjoy large public support, so civic society – if not the driver behind setting it up – should be informed regularly about developments. The selection process of the actual ombudsperson should be designed to guarantee broad support. Direct access for citizens to deliver inputs and receive information is important.
- Transparency: In order to enjoy and increase trust needs a clear mandate to access all information and early in the policy-making process. In return, it should maintain open relationships with all stakeholders during investigations and should report regularly about its work in a format that is accessible to all citizens.
- Authority: Research has shown that the 'shadow of enforcement' is very important for effective intervention, even if it is not used. The ombudsperson should have the opportunity to put actions or policies on hold if evidence on the long term consequences are insufficient. At best, he or she can call on a court if subsequent delivery of more information is not convincing, standing up for future generations.

Source: Göpel (2011)

National level

In any global agreement, success is determined by what happens at the national level and below. Do governments at all levels set up processes to implement? Without appropriate national laws and institutions, nothing will happen. Effective national environmental governance supports and complements efforts to improve international mechanisms for sustainable development. Effective national environmental governance helps ensure that parties to international environmental agreements actually enjoy the benefits that those agreements are supposed to provide. National processes which involve the general public are even more likely to be implemented. And sound national governance can allow global companies to operate in a similar way in all countries.

The World Resource Institute has developed a set of seven common principles that form a basis for effective environmental governance. These include the need that:

1   environmental laws should be clear, even-handed, implementable and enforceable;
2   environmental information should be shared with the public;
3   affected stakeholders should be afforded opportunities to participate in environmental decision-making;
4   environmental decision-makers, both public and private, should be accountable for their decisions;
5   roles and lines of authority for environmental protection should be clear, coordinated, and designed to produce efficient and non-duplicative programme delivery;
6   affected stakeholders should have access to fair and responsive dispute resolution procedures;
7   graft and corruption in environmental programme delivery can obstruct environmental protection and mask results and must be actively prevented.

These core principles enable countries to address environmental problems based on sound science (WRI, 2011).

Summary

This chapter has tried to identify achievable strategies to address the gaps and re-build strengths of the international and national governance structures that have built up over the last forty years. To summarize the recommendations that should be undertaken by 2015:

International

- create a World Environment Organization;
- create a Sustainable Development Council of the UN General Assembly to deal with new and emerging threats, which should replace the present UN Commission on Sustainable Development;
- create a high-level segment of ECOSOC to address implementation of Agenda 21, the JPoI and the results of Rio+20 through a non-policy forum including full participation of stakeholders;
- create a Sustainable Development Board to help deliver sustainable development at the 'one country programme' level;
- establish a High-Level Committee on Sustainable Development as the fourth committee of the UN Secretary-General's Chief Executive Board.

Regional

- create regional conventions for Principle 10 of the Rio Declaration.

National

- create national Ombudspersons for Future Generations.

Conclusion

We are committed to a strong and vibrant multilateralism that has at its heart the United Nations. Over the years in the area of economic, social and environmental affairs the UN's primary role has been of providing a global framework or context for actions that must be taken on *other* levels, regional, national or sectoral.

The challenges ahead are going to need a strong, well-resourced and dynamic UN that is capable of taking action in its own right. One which needs to be adequately funded, that is less fragmented, that is dynamic and able to deal with new challenges and crises firmly and quickly. This, frankly, will require some reduction in national sovereignty by individual countries for the sake of all the people living on our only one Earth.

The UN provides our best chance to all live together on this planet in a fair and equitable way. The question will be, are we clever enough to realize that in time?

Chapter 10

The implementation gap

Introduction

We know that Agenda 21 offered a practical approach to applying sustainable development policies at the local and national level, and the aspirational set of Principles that make up the Rio Declaration which gave us a set of guiding principles in the journey to a more sustainable planet. These successful agreements of the UNCED left an important and lasting legacy based on sustainable development needs and with a view on achieving more sustainable outcomes. But were they implemented?

This chapter draws from a project undertaken by Stakeholder Forum, with a core team of researchers familiar with the area of work, for the United Nations Department of Economic and Social Affairs (UN DESA). The project conducted a review of the status of implementation of both Agenda 21 and the Principles of the Rio Declaration that were agreed at the UN Conference on Environment and Development (UNCED) in 1992.

Agenda 21 Chapters – overview

Overall, progress on Agenda 21 has been limited. Despite being a comprehensive plan to deliver sustainable development, implementation has not been systemic. Of the 39 Agenda 21 Chapters, 31 were rated as having only made limited progress to date. An additional three Chapters were rated as having made no progress or regression. Only five Chapters were rated as achieving good progress and none were rated as fully achieved. The few good examples of where Agenda 21 has achieved positive and lasting outcomes include Chapter 7 of Agenda 21 which placed particular focus, with an obvious view to the future, on urban growth and sustainable urban planning. Chapter 16 has improved the regulation and use of biotechnology. Overall, progress on Agenda 21 implementation has been very poor.

The scorecards for both the Agenda 21 Chapters and the Rio Principles are based on an indicator rating system. This system uses indicator systems of four ratings to assess progress to date.

Sustainable Development 21 Scorecard Traffic Light Rating System

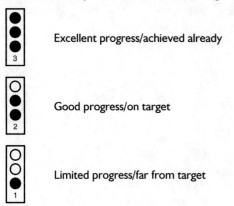

Excellent progress/achieved already

Good progress/on target

Limited progress/far from target

No progress or regression

The scorecard ratings are a subjective assessment based on the knowledge and expertise of the relevant author(s) of each section. These ratings also include a brief rationale to justify the rating.

SD21 Agenda 21 scorecard

Chapter 2 International cooperation to accelerate sustainable development in developing countries and related domestic policies

Assessor 1

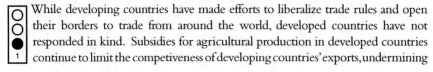

The efforts made by developing countries in terms of trade liberalization have not been matched by efforts from developed countries in terms of agricultural subsidies reductions. As such, the Doha Development Round has been in a stalemate for a long time. The amount of subsidies has reduced over the last two decades but organized schemes such as Aid-for-Trade are static. ODA is not enough nor as much as promised, plus the aid system has inherent problems leading to corruption and lack of devolution. A good number of countries have had debt relief but this is not enough both in number of countries and amount of relief.

Assessor 2

While developing countries have made efforts to liberalize trade rules and open their borders to trade from around the world, developed countries have not responded in kind. Subsidies for agricultural production in developed countries continue to limit the competiveness of developing countries' exports, undermining the supposed benefits of liberalization. Moreover, developing countries are falling behind in their aid commitments, coming nowhere near the 0.7% of GDP promised.

Final rating

Chapter 3 Combating poverty

Assessor 1

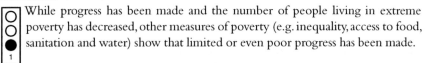

Significant progress was being made towards the MDG 1 of reducing the number of people living on less than US$1.25 per day but this has been seriously hindered by the financial crisis and as such the target is unlikely to be met. MDG 2 on increasing levels of education will not be met either with, only a slow increase in the number of children in school (including an increase in the number of girls). Child mortality has fallen but not as fast as expected, and women are still over-represented in the informal employment sector. The poverty gap has reduced overall but there are now more suffering from chronic hunger. This is also affected by the lack of ODA.

Assessor 2

While progress has been made and the number of people living in extreme poverty has decreased, other measures of poverty (e.g. inequality, access to food, sanitation and water) show that limited or even poor progress has been made.

Final rating

Chapter 4 Changing consumption patterns

Assessor 1

Unsustainable consumption patterns have continued to rise at a steady pace in industrialized countries but remain at an unsustainably high per capita plateau with very little evidence of reducing or any concerted efforts globally to address this problem. BRIC countries are seeing blooming consumer classes that aspire to high per capita consumption levels, and other developing countries will follow suit in time.

Assessor 2

Despite a number of initiatives and increasing levels of awareness and discussion surrounding SCP, the world has seen extremely little, if any, progress in regard to reaching the objectives outlined in Chapter 4. Since UNCED the world has seen a steady growth in consumption, and consumption patterns remain very high in certain parts of the world – with dramatic increases in the consumer population of India and China. Yet, the basic consumer needs of an even larger section of humanity are not being met. Whilst production systems have become more efficient, the patterns of consumption appear to have become more unsustainable; supported and exacerbated by the globalization of production and subsidies, and with very little in terms of national policies and strategies to encourage changes in unsustainable consumption patterns (a target outlined in the Chapter), globally, consumption has spiralled dramatically out of control. The ecological footprint of the global population has increased by over a third since the production of Agenda 21.

Final rating

Chapter 5 Demographic dynamics and sustainability

Assessor 1

There has been slow, but some positive progress with family planning and the use of contraception, but in key population growth areas, contraception levels remain low; however global fertility levels are reducing, which is helping the low contraception levels. There have been some successes in reducing infant mortality due to MDG motivation, but the target is far from reached in the majority of regions.

Assessor 2

The global focus on demographic dynamics has actually declined somewhat due to it becoming apparent that it is actually consumption rates which pose a greater threat to sustainable development. Largely through the work of intergovernmental institutions, there have been important steps forward in developing and disseminating knowledge and data on the links between demographics and sustainability. Yet large gaps still exist in our understanding of the relationship between these factors and the broader global environmental system. The creations of specific MDGs attempting to combat certain population-related development issues have had a positive impact in many developing countries and communities. Nonetheless, most nations remain significantly off course from achieving these targets by 2015. In general, therefore, it would appear that the formulation and implementation of integrated population-sustainable development policies remains absent at both the national and local levels.

Final rating

Chapter 6 *Protecting and promoting human health conditions*

Assessor 1

While progress has been made in reducing child and maternal mortality, it has not been enough; 1 in 4 children are still underweight and family planning funding has decreased. HIV/AIDS treatments are providing significant benefits but the number of new infections is outstripping the supply of treatment. Malaria has garnered increased attention but the impacts of this have not yet been felt and the area is still rife with inequalities between rich and poor. Diarrhoeal diseases are proving to be the biggest challenge with a lack of attention given to sanitation and water provision. The sanitation MDG is lagging the farthest behind. To meet the health MDGs there needs to be another US$20 billion injected. Environmental health hazards such as indoor cooking systems are being ignored but are having serious impacts on health of the poor, especially in urban areas.

Assessor 2

Progress has been made in some areas (infant mortality and communicable diseases), other areas still suffer from lack of progress (environmental causes) and health issues are still widespread and endemic.

Final rating

Chapter 7 *Promoting sustainable human settlement development*

Assessor 1

While there are a few examples of urban projects, the overall situation is one of continuing socio-economic inequality. The Right to Adequate Housing became a Human Right in 2006 and the proportion of the population living in slums has decreased, but in absolute numbers there are now many more people living in slums than previously. A lack of housing and the out pricing of the majority of the population from accessing adequate housing is a problem in both developing and developed countries. One major reason for the lack of settlement initiatives is the lack of funding going into this area. Furthermore, adequate water and sanitation provisions are a major part of suitable human settlements, yet they are the areas most lacking in progress. The main problem is that there has not been the required modernization within settlement planning that is needed to deal with the increased urbanization and population growth. Most benefits at the moment are being accrued by the richer members of society.

Assessor 2

While there are some good examples of progressive urban policy, the socio-economic inequalities and negative environmental issues within many urban areas remain widespread in both developing and developed countries. Slum populations are rising and conditions continue to worsen.

Final rating

Chapter 8 Integrating environment and development into decision-making

Assessor 1

Whilst most countries have created institutions and laws specifically aiming at mainstream environment and development in decision making processes, their influence and impact at the policy, planning and management levels remain limited in the majority of countries. Numerous market-based instruments and other incentives have emerged to promote the integration of environmental considerations into business practices. These have had a notable impact in some cases, however on the whole their scope and impact remain limited, with 'business as usual' prevailing in most regions, countries and communities. Despite advances in technology and the development of global mechanisms to support their implementation, most countries – especially in the developing world – do not possess fully functioning systems of integrated environment and economic accounting systems.

Assessor 2

There has been some implementation and integration of National Sustainable Development Strategies but far from complete coverage. UN agencies have done some work in advancing this agenda (e.g. IAP and PEI). While progress has been made using EIAs in Europe, this practice is limited elsewhere.

Final rating

Chapter 9 Protection of the atmosphere

Assessor 1

Progress in limiting the emission of greenhouse gases into the atmosphere has been non-existent, with annual CO_2 emissions growing year on year, and even the rate of growth increasing. Efforts to achieve international agreements on curbing emissions have repeatedly met with failure, with little to suggest concrete measures will be taken in the future. Separately, while emissions of ozone and particulate matter have decreased or stayed the same in developed regions, the picture is far less promising in developing countries with huge rises in emissions observed.

Assessor 2

While some progress has been made in this area (e.g. ozone depletion) greenhouse gas emissions and other atmospheric pollutants remain a huge problem and growing. Anthropogenic climate change is one of the biggest challenges to sustainability

Final rating

Chapter 10 Integrated approach to the planning and management of land resources

Assessor 1

There has been minor success in implementing the objectives outlined in Chapter 10 – the limited (due to lack of investment in human and fiscal resources) implementation of suitable land use and land management policies, strategies and action plans; the increase in international and regional initiatives and institutions; the quality and quantity of land-use information is improving through such technologies, which are being utilized alongside socioeconomic data to inform a comprehensive collection on land use. Yet, the pace of progress for the implementation of Chapter 10 of Agenda 21 remains uneven and inefficient with large unnecessary overlapping and conflicts in efforts at various government levels, the extremely alarming incidence of tenure insecurity and the scale of 'land grabs', and ineffective and weak dissemination of technologies and data provision (particularly at the national levels). Such issues are all compounded by the high levels of corruption among elites and by the increases in human population which will decrease the average availability of land per person globally.

Assessor 2

 UN's FAO has led various initiatives to promote sustainable land use (e.g. promotion and development of planning, management and evaluation systems for land and land resources; the development of land evaluation frameworks; and land-use databases). However, progress of further and more widespread implementation of such strategies remains limited due to lack of investment of human and financial resources.

Final rating

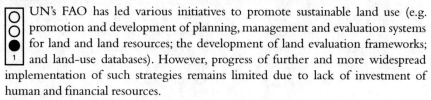

Chapter 11 Combating deforestation

Assessor 1

 Overall rates of deforestation have decreased thanks largely to widespread afforestation/reforestation programmes, however the destruction of primary forest remains alarmingly high across all regions. Progress in sustaining the multiple roles and functions of forests has therefore been limited. This can be attributed to the failure of many countries to effectively combat the drivers of deforestation – especially agriculture. Numerous institutions and initiatives have been created at the global, regional and national levels to improve the observation and systematic assessment of the full value of forests. Nonetheless, the impact of these advances and initiatives continue to be frequently undermined by the poor governance and weak institutions present in the developing countries which house the Earth's largest forest resources.

Assessor 2

While the last two decades have seen a significant increase in efforts to conserve biodiversity through forest initiatives, the FAO stresses that the current rate of deforestation is still 'alarmingly high'. In the last two decades the overall rate of deforestation has shown signs of decreasing. However, this is not due to decreased wood removal, but rather improved afforestation rates.

Final rating

Chapter 12 Managing fragile ecosystems: combating desertification and drought

Assessor 1

 Although some progress has been made to implement the objectives outlined in Chapter 12, such progress remains inadequate and is increasingly becoming hampered by the present and projected impacts of climate change, increasing global human population and increasing levels of consumption. Major obstructions to implementation remain the lack of practical and effective information and monitoring systems (particularly in relation to the socio-economic impacts of desertification and drought) and institutional inadequacies – particularly those of the UNCCD, which remains disjointed and unconnected from two additional UN conventions (UNCCC, UNCBD).

Assessor 2

The effectiveness of the United Nations Convention to Combat Desertification (UNCCD) has been limited, due to insufficient interaction with the scientific community and a lack of harmonization with the Conventions on Biodiversity and Climate Change. Moreover, while the Global Earth Observation System of Systems (GEOSS) might have strengthened information and monitoring at the global level, efforts at the regional and national levels have been less successful, with Africa in particular lacking the scientific capacity to adequately assess desertification. Most worryingly, there is deep concern over the capacity of developing countries to cope with drought induced by climate change.

Final rating

Chapter 13 Managing fragile ecosystems: sustainable mountain development

Assessor 1

Despite successes – for example, with the Mountain Agenda receiving increasing recognition and action across levels, a reasonable programme of work on mountain biological diversity, and successful implementation of various PES schemes (although not widespread) – Chapter 13 has failed to address a number of critical issues effectively (e.g. freshwater, biodiversity, cultural diversity and heritage, infrastructure development for mountain communities). There remains a significant dearth of comprehensive policies and laws, across all levels, to specifically protect mountain areas and communities, with mountain populations continuing to be marginalized within sustainable development policies.

There remains a considerable gap in terms of scientific knowledge and mountain-specific data to provide a higher level of understanding of mountain regions. Such specific scientific knowledge and data is critical when considering the impacts of global climate change. Finally, the promotion of alternative livelihoods has been meagre and has seen very little successful activity.

Assessor 2

The effort to strengthen the sustainable development of mountains has been significantly undermined by sector-based institutional structures which fail to account for cross-cutting issues and are insufficiently harmonized. Efforts to improve data collection and monitoring are similarly lacking, severely limiting the overall effectiveness of sustainable mountain development initiatives. The lack of comprehensive national mountain development strategies is a further aggravating factor.

Final rating

Chapter 14 Promoting sustainable agriculture and rural development

Assessor 1

Whilst we are still yet to see the 'major adjustments' called for by Chapter 14, a small level of progress has been made towards objectives outlined in the chapter, and the recent renewed international focus on agriculture as a mechanism for sustainable rural development is encouraging. Yet, with increasing levels of competition for land and other natural resources, higher energy prices, volatile food prices and new market demands (e.g. biofuels), combined with the lack of investment seen in agriculture over the last two decades, weak technology transfer, institutional incoherencies and weaknesses across all levels, poor infrastructure and lack of access to markets, much still remains to be achieved. Particularly so in light of the impacts of climate change, which are predicted to increase food insecurity and hamper rural development, especially within sub-Saharan Africa.

Assessor 2

Agricultural productivity has seen huge gains across the world, but the situation in sub-Saharan Africa continues to be bleak, with no increases in labour productivity. Moreover, while public investment in agriculture in Asia has risen, African governments have failed to live up to the commitments of the Maputo Declaration. Growing populations and resource scarcity look likely to hit the poorest countries hardest, while the ability of agricultural systems in developing countries to cope with the impact of climate change is also in question.

Final rating

Chapter 15 Conservation of biological diversity

Assessor 1

Efforts have been made at all levels to protect and preserve biodiversity: 170 countries have national biodiversity action plans; public awareness campaigns and scientific research and monitoring efforts have increased; and the number of protected areas globally has risen. But despite these efforts, in the 20 years since the Rio Summit, biological diversity has continued to decline and prognosis for biodiversity is grim with high levels of extinction expected to occur over the next hundred years. The underlying drivers of biodiversity loss, unsustainable use of biological resources, pollution, habitat destruction, invasive species, and climate change, continue to increase.

Assessor 2

Since the Rio Summit, biodiversity has continued to decline, and prospects for the future are bleak, with extinction likely for many species. None of the objectives of the Convention on Biological Diversity were met globally by 2010, with either no progress at all or regression in certain areas, such as unsustainable consumption of biological resources and protection of traditional knowledge. Moreover, overall levels of funding remain inadequate to achieve the necessary levels of biological conservation.

Final rating

Chapter 16 Environmentally sound management of biotechnology

Assessor 1

The biotechnology industry has seen huge growth over the past 20 years, yet the benefits for development, particularly in poorer countries as highlighted in Agenda 21, haven't been realised. There is still a huge amount of controversy surrounding many biotech applications, e.g. genetically modified crops and stem cell research, and there are countless regulations related to different biotech applications, which has led to incoherent and conflicting national and regional policies, further dividing opinion. Progress in international cooperation has also been slow, partly because of private sector dominance, but also these conflicting regulations and controversy.

Assessor 2

 Attempts to create enabling mechanisms for the development and the environmentally sound application of biotechnology have been largely piecemeal, with many examples of regional and international legislation but little in the way of a comprehensive, unifying framework. While biotechnology is growing in importance, its application in the developing world has been limited, with activity largely confined to industrialized countries. There is also a profound lack of consensus over the potential benefits and risks involved in biotechnology, considerably undermining public and political confidence.

Final rating

Chapter 17 Protection of the oceans, all kinds of seas, including enclosed and semi-enclosed seas, and coastal areas, and the protection, rational use and development of their living resources

Assessor 1

 In the 20 years since Rio, the state of the world's oceans and coastal areas has continued to decline. Coastal areas are being heavily degraded with about 400 zones now intermittently or always oxygen depleted, including over 200 dead zones; 50 per cent of global fish stocks are fully exploited with 40 per cent of total fish catch done unsustainably. Management of high seas fisheries is only in its infancy and small island developing states are still suffering from loss of biodiversity, habitat loss, coastal degradation, sea level rise and extreme weather events. Progress has been made in substituting integrated coastal zone management (ICZM) and ecosystem-based approaches for sectoral approaches; however, implementation has been difficult.

Assessor 2

 Significant progress at the global and regional level with development of governance and commitments to ICZM. However, as seen in many other areas, national and local implementation is slow or non-existent in many cases. The result being that marine ecosystem health continues to decline rapidly with most fisheries either in decline or over-exploited. Some success stories: EU Marine Strategy Framework Directive has paved the way for marine spatial planning and many countries are advancing on this. However, this is a long way from any targets and a long way from reversing the damage that is ongoing.

Final rating

Chapter 18 Protection of the quality and supply of freshwater resources: application of integrated approaches to the development, management and use of water resources

Assessor 1

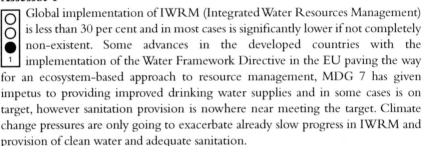

Global implementation of IWRM (Integrated Water Resources Management) is less than 30 per cent and in most cases is significantly lower if not completely non-existent. Some advances in the developed countries with the implementation of the Water Framework Directive in the EU paving the way for an ecosystem-based approach to resource management, MDG 7 has given impetus to providing improved drinking water supplies and in some cases is on target, however sanitation provision is nowhere near meeting the target. Climate change pressures are only going to exacerbate already slow progress in IWRM and provision of clean water and adequate sanitation.

Assessor 2

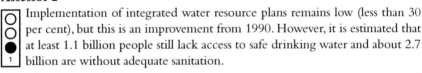

Implementation of integrated water resource plans remains low (less than 30 per cent), but this is an improvement from 1990. However, it is estimated that at least 1.1 billion people still lack access to safe drinking water and about 2.7 billion are without adequate sanitation.

Final rating

Chapter 19 Environmentally sound management of toxic chemicals, including prevention of illegal international traffic in toxic and dangerous products

Assessor 1

Good progress has been made in most of the programme areas outlined in Agenda 21. The Stockholm and Rotterdam Conventions, as well as the Strategic Approach to International Chemicals Management (SAICM) and the EU REACH legislation, are helping to improve chemicals management. Some issues such as illegal trafficking in chemicals remain a serious problem, and many countries, particularly developing countries, have a long way to go to improve their national frameworks for managing chemicals. Overall progress is encouraging though.

Assessor 2

 Globally numerous international institutions and initiatives have been created to deal with the management and regulation of chemicals. Successes include ozone-depleting substances, mercury and DDT. Labelling of chemicals has also seen significant progress. However, globally the growth of the chemicals industry is enormous, with production and consumption in developing countries increasing. There is a link between poverty and increased exposure to toxic chemicals. Despite progress, the consensus is that they are insufficient to achieve the goals set out in Agenda 21.

Final rating

Chapter 20 Environmentally sound management of hazardous wastes, including prevention of illegal international traffic in hazardous wastes

Assessor 1

 Although some regions have made significant progress, notably the EU, which has introduced much more stringent laws related to hazardous waste (WEEE directive, revised Waste Framework directive, etc.), globally hazardous waste generation continues to increase, and illegal trafficking, dumping and trans-boundary movements of waste (particularly WEEE) remain serious issues. The Basel Convention, which is more or less the only international legislation dealing with hazardous waste, has some serious weaknesses which need to be addressed.

Assessor 2

 Economic growth, industrialization and urbanization have led to rapid increases in volumes of hazardous waste, with electrical and electronic waste increasingly giving cause for concern. Efforts by the EU to deal with the volumes of hazardous waste currently produced are undermined by the inability of developing countries to do likewise, with companies from industrialized regions often paying poor countries to accept waste. Trans-boundary waste, including illegal trafficking, undermine the capacity of regulators to do anything about the problem, with initiative such as the Basel Convention seemingly having little effect.

Final rating

Chapter 21 Environmentally sound management of solid wastes and sewage-related issues

Assessor 1

 Same old story really: progress made in developed countries at waste minimization, however per capita waste levels are completely unsustainable. Annual waste production globally increasing by approximately 8 per cent and data too unreliable to know if progress is really being made or simply 'exported' either physically through dumping waste on other countries or indirectly through outsourcing waste-producing industries while still reaping the benefits. Some success stories though, mainly in developed countries, with significant increases in recycling rates and innovative new technologies for reuse, and new regulations to put final disposal burden onto the producer aims to promote more sustainable eco-design – however key issue of reducing waste production altogether still not being addressed.

Assessor 2

Significant progress has been made in some areas of solid waste management, particularly recycling and final disposal. The majority of this progress has taken place in developed countries though, with developing countries often lacking the resources to cope with increasing volumes of waste, which is so closely linked to economic development. Minimizing waste continues to be the most challenging aspect of waste management, but this has in the last few years, become the focus of many countries' national strategies, and positive results are beginning to be seen in developed countries.

Final rating

Chapter 22 Safe and environmentally sound management of radioactive wastes

Assessor 1

The International Atomic Energy Agency has introduced several laws, regulations and codes of practice over the past 20 years that have greatly improved the management of radioactive waste. Some of the specific targets set out in Agenda 21 (e.g. restrictions on the trans-boundary movements of waste, and banning dumping at sea) have already been achieved, and overall progress is encouraging. There are still issues to be addressed though, e.g. high level waste disposal, and the IAEA's Joint Convention needs strengthening as it does have its flaws.

Assessor 2

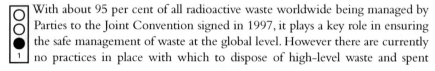 With about 95 per cent of all radioactive waste worldwide being managed by Parties to the Joint Convention signed in 1997, it plays a key role in ensuring the safe management of waste at the global level. However there are currently no practices in place with which to dispose of high-level waste and spent

nuclear fuel. The general consensus is that deep geological disposal is the best option for high-level waste, and underground facilities are currently in the planning process. This is expensive and highly technical, and some countries lack capacity to implement such disposal practices. Additionally, legacy waste which has been poorly disposed of remains a huge environmental and health risk.

Final rating

Chapter 23 Major Groups – preamble

Assessor 1

 Involvement of multiple social groups in governance processes has increased significantly on all spatial levels since 1992. It is widely recognized that broad public participation in decision-making is a prerequisite for sustainable development, and new forms of participation have emerged. The Internet and new information and communication technologies have revolutionized access to information. Still there is need for improvement, since not all governments are equally eager to involve their citizens in meaningful partnerships. Only a few countries have institutionalized constant participation of Major Groups in national decision-making for sustainable development. Closer collaboration is needed for transparency, legitimacy and accountability.

Assessor 2

The quantity of non-state actors engaged in UN summits and processes has grown tremendously since the adoption of Agenda 21. Agenda 21's establishment of the concept of nine Major Groups has increased the diversity of actors involved in many UN processes. In order to not go against the UN Charter, the inclusion of all Major Groups in the UN system needs to be engineered through organizational NGO constructs. However, there are still occasions when governments are meeting and Major Groups are not included, and other cases when they are allowed token presence in negotiations but there is a lack of meaningful participation. Only a few countries have institutionalized constant participation of Major Groups in national decision-making for sustainable development.

Final rating

Chapter 24 Major Groups – women

Assessor 1

 General awareness and tracking of gender issues has increased, but the man is still the norm, and the overall situation for the world's women is far from target. Although differences are big across regions, women remain the poorest of the poor everywhere. Many women have experienced a decline in their quality of life and a number of governments have turned back advances in women's autonomy. Global success stories are improvements in literacy and education for girls and women, the ratification by most governments of international women's rights treaties, and the right for women in most countries to hold public office.

Assessor 2

The women's Major Group has been very successful in their overall participation in global processes, and their activities clearly predate 1992 and Agenda 21. It seems obvious that their successes are related to areas that traditionally have been labelled women's issues – health, population welfare – but Agenda 21 also gave women a clear role in sustainable development. Gender mainstreaming is now a household word in all activities, and there is at least on the surface little resistance to giving women a role. Still this acceptance is often symbolic, and pertains to some countries more than other.

Final rating

Chapter 25 Major Groups – children and youth

Assessor 1

Young people still face disproportionate levels of poverty, gender discrimination and health problems, and global youth unemployment recently hit a high record. On the positive side, the world has seen growing acceptance of youth as legitimate actors in decision making since UNCED, and the UN has put in place support structures for promoting the role of youth. The Convention on the Rights of the Child has become almost universally ratified, though implementation levels vary. Many governments recognize the need to invest in the young generation for sustainable development and are committed to creating improved opportunities in the coming years.

Assessor 2

Integrating the work of youth in the UN is problematic as youth is a transitional group. As it is still neither well understood nor integrated into institutional systems or processes, the group is often paid lip service to, and its presence often becomes symbolic – 'it is good to be seen with youth'. Integrating youth into negotiations is improving, and would be nowhere had it not been for the recognition by Agenda 21, but progress has been difficult, uneven and slow. Trying to integrate children in the work of the UN in general and negotiations in particular, also shows a particular lack of understanding for how this particular group lives and operates. Involving them in negotiations – as they are found at CSD – would be a waste, as these processes are meant for adult professional negotiations. To work with children, organizations need to be developed with that in mind and run by people with special knowledge. The UN does not possess this at the moment. Mentoring programmes for youth concerning negotiations, process understanding, etc., have been on the agenda but are still not developed and there is a huge lack of interest in doing this. Consequently there is a long way to go before the children and youth group has a proper position within the UN.

Final rating

Chapter 26 Major Groups – indigenous peoples

Assessor 1

Targeted initiatives and international support structures for indigenous peoples have been established since UNCED, including the UN Declaration on the Rights of Indigenous Peoples. Implementation on the national level has been uneven, and while improvements are visible in some places, many states do not even acknowledge the presence of indigenous peoples in their countries. There is a long way to go for improving the living conditions of indigenous peoples, who are still marginalized and experiencing more poverty and health problems than the rest of the population across many regions. Traditional knowledge and cultural lands are too often disrespected.

Assessor 2

UNCED in Rio in 1992 was clearly a breakthrough for indigenous peoples in many aspects, not least with the recognition of the 's' in peoples. It is also obvious that Rio started a process that gave the indigenous groups an opportunity to pursue their policies for representing their peoples and consequently a recognition of participation in process globally. Their rights were respected, and their numbers in global UN and UN-related processes grew. Still, their participation is hampered by a general lack of resources, and they are far from reaching their targets.

Final rating

Chapter 27 Major Groups – NGOs

Assessor 1

The status and importance of NGOs has increased tremendously over the last decades. NGOs play roles as moral stakeholders, watchdogs, mediators, implementers, lobbyists, and experts. They have become increasingly professionalized and UN agencies have grown dependent on NGOs in mutually beneficial relationships. Multiple NGO networks are spearheading different aspects of sustainable development. On the national level, NGOs have in some cases become the main service providers towards sustainable development by taking over responsibilities that would normally be the task of governments. Most governments are in dialogue with NGOs and encourage their initiatives, while other governments are still suppressing NGOs.

Assessor 2

Rio+20 was a breakthrough for civil society and work on international processes. The number of registered NGOs with the UN – all divisions soared after 1992. It would be impossible as well as incorrect to relate the success in number to the Rio process alone. There is a drive inherent in the nature of NGOs that would always make sure the sheer number of NGOs would grow as well as their related activities.

There has been considerable success in integrating NGOs in the work of all entities of the UN, process as well as implementation, and quite clearly the NGOs have been driving process and implementation at all levels in the work on sustainable development, local, national, regional and global. Still the process is organic and ever evolving, thus targets are not fulfilled by any means.

Final rating

Chapter 28 Major Groups – local authorities

Assessor 1

Local Agenda 21 has been one of the most extensive follow-up programmes to UNCED and is widely cited as an unprecedented success in linking global goals to local action. Almost all local authorities around the world have adopted some kind of policy or undertaken activities for sustainable development, either as a main priority or as a cross-cutting issue. The excellent progress so far does not mean that the work is over, but rather that there is potential to build further on the success. Multi-level governance is needed, as well as increased integration between local authorities and multi-stakeholders in their communities.

Assessor 2

The foundation of ICLEI in New York in 1990, or the 'International Council for Local Environmental Initiatives' as it was first called, heralded global interest among local authorities for sustainable development. ICLEI is today one of several global organizations consisting of local authorities as members, and these are proofs of interest among local authorities to work on sustainable development. The initial high level of activity seems to have waned on all levels among local communities, and ICLEI did show during a few years less interest in active participation in UN processes directly related to sustainable development, and did prioritize other environmental concerns such as climate, biodiversity etc. Constantly revising its goals, and questioning the position local authorities hold in UN processes, they do not feel they have found a relevant position in these processes. Despite these issues and a few setbacks, a degree of recognition to the success of many local authorities must be given, even if there are too many municipalities not participating.

Final rating

Chapter 29 Major Groups – workers and trade unions

Assessor 1

Most global trends for workers have gone in the wrong direction since UNCED. Income inequalities have grown dramatically in most regions of the world and are expected to rise further, and unemployment rates are the highest ever reported. Work-related accidents, injuries and deaths are unacceptably common and have been increasing in developing countries. Changes in workforce structures have made new kinds of occupational health problems common. Workers' conditions are dependent on national legislation and vary between countries. Companies seldom take measures beyond the minimum required to improve the life situation for their workers. Trade unions are often threatened.

Assessor 2

Being the oldest non-governmental entity in the UN family and used to being a serious and negotiating member of the international community – the tripartite agreement with ILO dates back to 1919, the commitment of trade unions to international process work cannot be attributed to Agenda 21 and Rio in 1992. Slow in accepting sustainable development as an issue, often fearing that sustainable development concerns might jeopardize job opportunities, trade unions have changed dramatically, and changed because of Agenda 21 and the ensuing work of the UN on sustainable development and environment-related work. Trade unions have shown great innovative skills in dealing with especially two of the three pillars of sustainable development – the social and economic ones – but struggling to find its proper role in relation to the many challenges inherent in the work of sustainable development, they have yet to feel totally comfortable in all aspects related to Agenda 21 work.

Final rating

Chapter 30 Major Groups – business and industry

Assessor 1

The private sector has potential to become a positive driver for sustainable development. Positive initiatives are emerging but are far too limited, such as social and sustainable entrepreneurship, green innovations and cooperative enterprises. Current global trends show that the vast majority of businesses prioritize short-term economic gains on the expense of social and environmental conditions. Environmental costs have grown with globalization of markets and industrial production patterns. Companies continue to violate human rights, exploit natural resources and pollute the environment. The business sector engages in green-washing, controls some areas of science, and lobbies hard to defeat regulation efforts.

Assessor 2

The business community seems always to be referred to as a must in talking about future development of the world, and more often than not because of the amount of money and finances it represents. The business community was a reluctant participant in Rio in 1992, but because of Rio and the ensuing work on environment, business has become an interested partner in sustainable development projects. It often represents a reactive force, and has at times acted in a more conservative manner than was necessary; still the business community has entered the sustainable development thinking with strong force. Despite laudable efforts of the local, national and global business communities to engage and fulfil the targets of Rio and Johannesburg, with a few notable exceptions, the business community is still underperforming.

Final rating

Chapter 31   Major Groups – science and technology

Assessor 1

The process of sound scientific knowledge production has improved as science has become more interdisciplinary and transdisciplinary. The field of sustainability science has grown rapidly and multiple research initiatives advance knowledge about Agenda 21 issues. Scientific global assessments have become common tools for improving communication and cooperation among the scientific and technological community, decision makers and the public. Codes of practice and guidelines related to science and technology are under development. The technological community provides innovation for sustainable development, but research engineers need to be reallocated to life-supporting sectors, since the majority is engaged in the military industry.

Assessor 2

Sustainable development has a number of times been labelled an impossible political concept and an equally impossible scientific concept. Were the science community to be evaluated only in terms of its presence and contribution to the CSD process this would indicate little progress. The main problems here are, as in other instances, that so much of CSD is politics. And scientists by nature, to keep and preserve their independent and objective position and role, shy away from politics. But as sustainable development issues have penetrated many other institutions of the UN where scientists are operative, the colour of appreciation changes. UNESCO, UNRISD, UN University – there are numerous institutions working with sustainable development related issues where scientists are active. UNEP being responsible for the environment pillar of sustainable development is a science based institution. Most of the global, regional and bilateral environment conventions have a basis in scientific facts backed up by scientific institutions, with UNFCCC and IPCC the best known of all. But targets listed in Agenda 21 and also reiterated in the JPoI have not been fulfilled.

Final rating

Chapter 32   Major Groups – farmers

Assessor 1

Food producers in rural communities in developing countries often live in poverty, even though their farming practices are low-resource and sustainable, which Chapter 32 aims to promote and encourage. On the other hand, large subventions are still provided to unsustainable high-resource agriculture, which is the largest single cause behind climate change and loss of ecosystem services. The amount of organic farming has grown in all world regions since UNCED, but constitutes only 0.9 per cent of the total agricultural land. Agricultural data and information has become more commonly available, but farmers are not sufficiently involved in the research-technology-knowledge nexus.

Assessor 2

Recognizing the importance of food and agriculture, FAO was one of the first of the UN specialized agencies to be established. The issue of food was also one of the hot topics for discussion in Rio in 1992, and provided the background for the development of one of the three Rio Conventions, the UN CBD. Rio+20 cannot claim the involvement of farmer organizations in global farming issues, but Rio brought sustainable farming to the global agenda. Inspired by Agenda 21 and the later development of the CBD and its protocol, including differentiated development of activities in other food and sustainable development-oriented agricultural issues, the role of farmers, and in particular the role of small farmers has been recognized and these groups have also been given an arena within which to act. Still, new issues keep emerging such as food safety and food security as well as bio-engineering, water shortages, etc.

Final rating

Chapter 33 Financial resources and mechanisms

Assessor 1

While Chapter 33 adequately lists resources and mechanisms vital to the implementation of Agenda 21, none of the chapter's financing methods were expanded upon enough to be effectively implemented. Furthermore, the absence of clear reporting procedures made the inadequate provision of Agenda 21 financing difficult to address. Nonetheless, at present there are increased resources available for sustainable development. Funding has steadily increased from Multilateral Development Banks and the Global Environmental Facility, and while ODA substantially fell following UNCED, development assistance levels have bounced back. Innovative financing methods have also grown in importance and possibility (i.e. Kyoto Protocol, High-Level Advisory Group on Climate Change Financing). However, if the implementation and measurement of sustainable development financing remains as vague as was set out in Chapter 33, these financing increases will become ill-used and unsustainable.

Assessor 2

 While funding has improved in recent years, funding arrangements and transfers of technology from developed to developing nations around the Agenda 21 outcomes have been not delivered as promised. ODA fell from US$62.4 billion in 1992 to US$48.7 billion in 1997. It was not until 2002 that it again topped the US$60 billion mark. This 'lost decade' was marked by regression of key development statistics with many of the world's poorest countries suffering from worsening poverty. However, aid flows from donor countries totalled US$129 billion in 2010, the highest level ever, and an increase of 6.5 per cent over 2009. Other challenges include inadequate measurement and reporting; lack of collaboration; questions of aid effectiveness; and trade and debt relief inequalities.

Final rating

Chapter 34 Transfer of environmentally sound technology, cooperation and capacity-building

Assessor 1

Although a raft of measures have been put in place to facilitate technology transfer, progress has generally been perceived as slow, with the rate of technology transfer having fallen over the lifetime of the Clean Development Mechanism. Policymakers have so far failed to deal with the complexity involved in transferring environmentally sound technologies from one institutional context to another, undermining the capacity of developing countries to benefit from 'leapfrogging'.

Assessor 2

Knowledge sharing has improved with the establishment of a multitude of partnerships and networks. Various initiatives exist to facilitate technology transfer. However, progress in actually transferring technology remains slow.

Final rating

Chapter 35 Science for sustainable development

Assessor 1

Since 1992 virtually, countries have strengthened the scientific basis for sustainable management, often through the creation of specific science-development institutions. Advances in the BRICS countries have been particularly pronounced, seeing the capacity and capability of the developing world becoming more closely aligned with that of their Northern counterparts. This, along with the continued development of specialized global scientific organizations – chief amongst them the IPCC – has resulted in a greatly increased understanding of global environmental processes. This is in turn closely related to tangible improvements made in long-term scientific assessment at the global, national and regional levels. There nonetheless remain significant problems surrounding the coherence of global scientific efforts and the myriad agencies which undertake both research and assessment. Despite some progress, the sustainable energy puzzle remains largely unsolved and many developing countries still lack the institutional capacities to place science at the centre of sustainable development programmes.

Assessor 2

Since the beginning of the twenty-first century, global investment in science and technology research and development has nearly developed, and scientific understanding of the Earth's carrying capacity and impacts of human activity has deepened considerably. A great many initiatives have improved the ability of countries around the world to appropriately assess progress in meeting sustainable development criteria, and various mechanisms have, to an extent, worked to incorporate scientific information into the decision-making process. That said, at the global level scientific assessment remains somewhat incoherent, and capacity constraints continue to impact upon a great many developing countries.

Final rating

Chapter 36 Promoting education, public awareness and training

Assessor 1

Achieving universal basic education and the eradication of illiteracy, central to Chapter 36 of Agenda 21, the Millennium Development Goals and the Education for All agenda, remains a distant dream, with 67 million children out of school in 2008 and 17 per cent of adults lacking basic literacy skills. Progress on re-orientating national education strategies towards sustainable development has been more promising, with a great many countries incorporating principles of sustainable development into curricula and establishing national coordinating bodies for the promotion of education for sustainable development. However, education for sustainable development lacks a clear definition, and whilst the outlook for education in general remains bleak, the capacity of education to act as an instrument for sustainable development appears to be limited.

Assessor 2

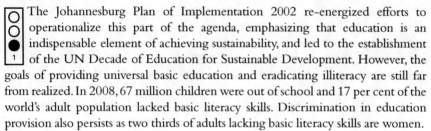 The Johannesburg Plan of Implementation 2002 re-energized efforts to operationalize this part of the agenda, emphasizing that education is an indispensable element of achieving sustainability, and led to the establishment of the UN Decade of Education for Sustainable Development. However, the goals of providing universal basic education and eradicating illiteracy are still far from realized. In 2008, 67 million children were out of school and 17 per cent of the world's adult population lacked basic literacy skills. Discrimination in education provision also persists as two thirds of adults lacking basic literacy skills are women.

Final rating

Chapter 37 National mechanisms and international cooperation for capacity-building in developing countries

Assessor 1

There are lots of national and international strategies for increasing countries' capacity but ultimately these are contradictory in theory and practice with most culminating in top-down generic solutions. What is needed is flexibility and iteration to fit within different contexts and to empower each country individually. The policies are too focused on measurable results when what is needed is endogenous change. However, there are a number of declarations and initiatives that are leading the way in this area, i.e. Paris Declaration and Accra Agenda for Action, plus signs of international cooperation with programmes such as the Delivering As One.

Assessor 2

Strategies for Sustainable Development (NSSDs) and poverty reduction strategies (PRSs) have emerged as the key mechanisms through which countries are able to assess their capacity needs and target improvements. However, capacity development has all too frequently been viewed as a technical, universal applicable process, and has tended to ignore the ways in which national capacity is a function of the local institutional and socioeconomic context. The results-driven perspective of developed countries, whose aid is frequently contingent on achieving strictly measureable objectives, has worked to undermine the long-term sustainability of capacity development initiatives.

Final rating

Chapter 38 International institutional arrangements

Assessor 1

Chapter 38 is rated as 'achieved already' since international institutional arrangements have been put in place as suggested in the chapter, and all mentioned UN agencies have made efforts to fulfil the roles envisioned for them in Agenda 21. However, the arrangements are not ideal since they include overlapping mandates resulting from a process of negotiation and compromise. Experience shows that the institutional support structure is not coherent enough for effective and efficient implementation. Greater coherence and institutional connections between different spatial levels are needed, and there is an urgent need to reform the institutional framework for sustainable development.

Assessor 2

The changes to the UN proposed in Agenda 21 have each come to fruition, and much has been done to unite the development and environment agendas at the international level. The establishment of the Commission on Sustainable Development represents a particularly significant achievement, given the complexity of the discussions at Rio. That said, there is concern over the ability of the CSD to live up to its mandate, and other institutional challenges remain, for example in the case of the frequently overlapping and contradictory Multilateral Environmental Agreements (MEAs). The lack of implementation apparent across a number of Agenda 21's objectives also brings the effectiveness of the UN's overarching structure into question.

Final rating

Chapter 39 International legal instruments and mechanisms

Assessor 1

Agenda 21 has been a significant catalyst for the generation and application of legally binding agreements in the environment and development domains. Multilateral environmental agreements have reporting requirements. The CSD provides review, assessment and fields of action in international law for sustainable development. UNEP and others have contributed to further development of implementation mechanisms. Effective participation in international law-making is supported by capacity-building services, training materials, and funding to developing country delegates to attend negotiations. An international dispute resolution mechanism purely for environmental or sustainable development issues is lacking. Multilateral environmental agreements would need to be clustered for coherence.

Assessor 2

A great many Multilateral Environmental Agreements (MEAs) have been negotiated since 1992, creating a legal mechanism which situates environmental issues in the context of sustainable development. Moreover, the United Nations Framework Convention on Climate Change (UNFCCC), the Convention on Biological Diversity (CBD), and the United Nations Convention to Combat Desertification (UNCCD) have each gone some way to strengthening the legal framework for sustainable development, successfully balancing the interests of different parties. However, it remains the case that legal commitments are often not matched by implementation, and there are also instances in which the legal framework is self-contradictory, as in the case of MEAs.

Final rating

Chapter 40 Information for decision-making

Assessor 1

Much has been done to strengthen frameworks of sustainable development indicators and provide a new basis for decision-making; the UN, OECD and EU have all worked to ensure environmental indicators are amenable to the demands of policymakers. However, the reduced capacity of developing countries to collect and analyse sustainable development data continues to give cause for concern, limiting the effectiveness of measures taken by the international agencies to harmonize environmental data at the global level. Insofar as bridging the data gap between developed and developing countries was a central objective of Agenda 21, it is far from clear that progress has been sufficient.

Assessor 2

While a great deal of effort has been put into developing and implementing sustainable development indicators, data collection and analysis remain a challenge, particularly in developing countries. Even where data exists, its reliability and quality is at times questionable. Enhancing countries' institutional capacity to collect and assess data remains a priority. Furthermore, global indicator frameworks, in seeking to harmonize environmental data sets at the international level, risks distorting the local picture and compromising traditional and indigenous knowledge.

Final rating

Successes

Arguably, Agenda 21's biggest success has come through driving ambition on what sustainable outcomes are achievable on a sector-by-sector basis. For example, our understanding of biodiversity, the contribution agriculture makes to development or the role of indigenous peoples in society has been advanced in no small part through Agenda 21.

Agenda 21 (and the original Rio Earth Summit more generally) brought the concept of sustainable development into common parlance if not making it a household phrase. It had a strong influence on the language of subsequent international agreements and documents (such as WTO preamble, the Cairo Agenda on Population (1994), the Social Summit outcome (1995), the Beijing Women's Conference (1995), the Habitat Agenda (1996), and the Rome Food Summit (1996). Overall, one clear and positive impact of Agenda 21 has been to help put the concept of sustainable human development at the heart of development, as opposed to more 'techno' solutions in the 'development decades' of the 1960s and 1970s (for example, strategies based on rapid industrialization and large-scale agricultural projects). There is, however, significant work to do before current development solutions can be considered to be a true rights-based approach to dealing with welfare, well-being and environmental issues. Such an approach would put people at the heart of development that is also sustainable.

Agenda 21 tried to address the issue of integration of environment and development through the creation of the Commission on Sustainable Development (CSD). The Commission was a compromise between those who wanted to transform the Trusteeship Council into a Sustainable Development Council, therefore making it one of the permanent bodies of the UN, and those countries that wanted no follow-up mechanism. The placing of the UNCSD as a functioning commission of the Economic and Social Council (ECOSOC) did have some early successes with the issues of Persistent Organic Pollutants (resulting in the Stockholm Convention), Prior Informed Consent (resulting in the Rotterdam Convention), oceans (the United Nations Open-ended Informal Consultative Process) and forests (UN Forum on Forests). It would initiate the conversations or key other processes and then set them off to be negotiated more formally in other fora.

Furthermore, Agenda 21 has engendered a much stronger notion of participation in decision making. This affirmation of the important role of non-governmental actors has percolated all levels of government, international law and international governance. This includes promoting a greater granularity in demographics for analysis and decisions. For example, Agenda 21 helped bring the gender dimension in all development work and beyond, including gender-differentiated official statistics.

The participation of the Major Groups – as outlined in Chapter 23 – has been improved with a formalized process in place to acknowledge the contribution to dialogues on sustainable development. Specifically, the status and importance of NGOs – as outlined in Chapter 27 – has increased tremendously over the last decades. NGOs play roles as moral stakeholders, watchdogs, mediators, implementers, advocates, and experts. They have become increasingly professionalized and UN agencies have grown dependent on NGOs in mutually beneficial relationships. Multiple NGO networks are spearheading different aspects of sustainable development. However, how much of this 'improved participation' is simply rhetoric is debateable and requires further exploration.

Local Agenda 21 has been one of the most extensive follow-up programmes to UNCED and is widely cited as an unprecedented success in linking global goals to local action. In 2002, over 6,000 local authorities around the world – the Major Group addressed in Chapter 28 – were found to have adopted some kind of policy or undertaken activities for sustainable development, either as a main priority or as a cross-cutting issue. This was often under the rubric of 'local Agenda 21'. The excellent progress so far does not mean that the work is over, but rather that there is potential to build further on the success. Multi-level governance is needed, as well as increased integration between local authorities and multi-stakeholders in their communities.

Agenda 21 represented a progressive vision for action that set standards of ambition and success incomparably higher than the plans of old in addition to cost. It also built a strong narrative for action which was in itself progressive.

Agenda 21 remains as relevant today as it was 20 years ago. While there are some gaps in coverage (see below for details), the issues that humanity is struggling with now are more or less similar to those covered by the chapters of Agenda 21.

Challenges

In retrospect the sectoral approach of Agenda 21 may have undermined the concept of sustainable development as one which encourages integration and cross-sectoral solutions. Viewing these issues in isolation has often paved the road for turf wars and 'silo-ization', both at the international level and at the national level. Often stretching the boundaries of a discussion to explore the inter-linkages of other sectors is either competition for attention or resources, or worse as a direct threat. Hence, different, but related topics are frequently treated in various fora with no link being established to connecting issues, generating policy incoherence and confusion. This has led to strategic gaming with interlinked issues being seen as trade-offs (e.g. trade vs. Intellectual Property Rights (IPRs) in food and biodiversity), and has generated policy incoherence and confusion. The UN's agencies have struggled to effectively address these inter-linkages.

Another issue is that some sectors were not included. This broke the all-encompassing nature of the document. For example, energy and mining are key sectors that were not included as Agenda 21 chapters. Moreover, key issues would today be more prominent than their space in Agenda 21, for example transport and waste flows. However, energy, transport and tourism were each discussed in 1997 in a five-year review from Rio.

Some areas of Agenda 21 have remained largely unsuccessful and could even be deemed failures. For example, sustainable consumption and production remain unsustainable in absolute terms. Although resource use has significantly reduced per unit of global economic output over the last 25 years (Jackson, 2009) (by around 30 per cent), globally we are using around 50 per cent more natural resources than we were over the same time period. Furthermore, this consumption of resources is distributed inequitably. North American per capita consumption is around 90 kg of resources per day, around 45 kg per day for Europeans and around 10 kg per day for people in Africa (Friends of the Earth, 2009).

Despite a number of initiatives and increasing levels of awareness and discussion surrounding sustainable consumption and production (SCP), the world has seen extremely little, if any, progress in regard to reaching the objectives outlined in Chapter 4 of Agenda 21. The ecological footprint of the global population has increased by over a third since the production of Agenda 21. Since UNCED, the world has seen a steady growth in consumption, and consumption patterns remain very high in certain parts of the world – with dramatic increases in the consumer population of India and China. Yet, the basic needs of an even larger section of humanity are not being met. Whilst production systems have become more efficient, the patterns of consumption appear to have become more unsustainable, supported and exacerbated by the globalization of production, and with very little in terms of national policies and strategies to encourage changes in unsustainable consumption patterns (a target outlined in Chapter 4); globally, consumption has spiralled dramatically out of control.

While some progress has been made around Chapter 9 – the protection of the atmosphere – (e.g. ozone depletion) greenhouse gas emissions and other atmospheric pollutants remain a huge problem and growing. Anthropogenic climate change is one of the biggest challenges to sustainability.

Chapter 7 – human settlement development – lacks meaningful progress. While there are some good examples of progressive urban policy, the socio-economic inequalities and negative environmental issues within many urban areas remain widespread in both developing and developed countries. Slum populations are rising and conditions continue to worsen.

In retrospect, Agenda 21 reflected a somewhat static view of the world. This was largely due to the fact that the agenda was cut into 40 sectoral pieces/chapters. Agenda 21 did not address the interconnectedness of the various goals, because it was not allowed to examine the economic system itself. Nor did it explore the fundamental drivers of sectoral and inter-country outcomes, which include:

- the role of corporations, and multinational corporations (MNCs) in particular;

- the role and impacts of trade and globalization;
- the role of international economic governance in helping steer the whole system; and
- the importance given to future generations in everyday policy-making.

Agenda 21 also failed to adequately address the institutional side. It underestimated the inertia and resistance of institutional structures at all levels. These issues included silo-ization, bias against developed country representation in rule-making, focus of politicians on 'development first' and a disconnect between different levels of government.

Furthermore, Agenda 21 had an implicit framework for action relying on nation states acting on their own for delivery, with some top-level international coordination. Agenda 21 was costed out at US$625 billion a year as developed countries sought to address their own unsustainable development patterns. It also had meant to create a doubling of overseas development aid (ODA) to US$125 billion a year after Rio, which in fact decreased. Agenda 21 was meant to benefit from the 'peace dividend' from the break-up of the Soviet Union so the figures were not thought to be great.

One of the impacts of the new CSD work programme in 2003 was the removal of the reviews of the implementation of Agenda 21 against ODA contributions. This resulted very quickly in a reduction of the level of participation by the donor government's aid divisions. In the first ten years, high-level involvement, even in some cases attendance by development ministers, kept the hope that funds would flow to deliver Agenda 21 as had been promised in 1992.

There had been an attempt by the UN Centre for Transnational Corporations to table a 41st Agenda 21 chapter on 'Transnational Corporations and Sustainable Development'. This was rejected and within two years the Centre had been closed down with its function shifted to United Nations Conference on Trade and Development (UNCTAD).

Trade had played only a small role in Rio. This issue was subsequently put to the WTO by the CSD as a challenge to the new body to integrate sustainable development into trade decisions. The WTO's founding agreement recognizes sustainable development as a central principle, but this is rarely a reality. For example, the promotion of subsidies that contribute to unsustainable development have been promoted under the WTO and are having impacts on the use of natural resources that are detrimental to the creation of a sustainable pathway of the planet. We still have unresolved the Doha development round ten years after it had started to be negotiated.

The main global economic institutions – the IMF and World Bank – have not meaningfully reformed their practices to embrace sustainable development. Although certain policies can be shown to support sustainability, and individuals within the Bank and Fund are promoting sustainability ideas, the overall activities of both institutions and the regional development banks have supported the present economic model.

The creation of the Interagency Committee on Sustainable Development (IACSD) to oversee the Task Managers for Agenda 21 did achieve some coordination and implementation in the UN agencies and programmes. But with additional funding this could have achieved much greater levels of implementation. The closure of the IACSD as a part of the UN reforms in the late 1990s reduced the coordination and integration amongst UN Agencies and Programmes, and the subsequent impact was felt.

Agenda 21 also failed to adequately address the institutional structures. It underestimated the inertia and resistance of institutional structures at all levels. These issues included silo-ization, bias against developed country representation in rule-making, focus of politicians on 'development firs' and a disconnect between different levels of government.

Conclusions

Agenda 21 retains strong relevance 20 years after the Earth Summit. And while the implementation of Agenda 21 has acquired considerable coverage amongst nation states, this remains far from universal or effective. At the end of Rio there was a perceived agreement that funding, capacity building and technology transfer would be forthcoming once developed countries moved out of recession. What was seen as the 'peace dividend' from the fall of the Soviet Union was where funding would come from. This did not materialize. Progress has been patchy, and despite some elements of good practice most Agenda 21 outcomes have still not been realized. Disputes continue on how to implement Agenda 21. For example, developed nations favour financing it through bilateral, regional and multilateral mechanisms and more and more through foreign direct investment. A path promoted in the 1990s after Rio was shown to mostly benefit nine countries. The Group of 77 developing countries still favour the implementation of the financial agreement in Rio and this now would include a separate, specific global fund, as well as commitments that financing will not be obtained through reallocation of existing development assistance.

Rio Principles – overview

The review of the Rio Principles shows that many of the principles have been transposed into further international laws or national instruments, but have not necessarily filtered down into meaningful action in practice. Without full compliance and enforcement mechanisms there is little to ensure that states comply with the objective and aspiration of the principles. However, there are some successes, such as Principle 10 (Access to Environmental Information) as enshrined in the Aarhus Convention which covers most European Union (EU) members.

Overall, progress on the Rio Principles has been very poor. Of the 27 Rio Principles, 18 (67 per cent) were rated as amber, meaning that only limited progress has been made to date. An additional 6 Principles (22 per cent) were rated red (no progress or regression). Only 2 Principles (7 per cent) were rated green (good progress) and 1 (4 per cent) was rated blue (excellent progress/full achieved). The summary scorecard on the implementation of the Rio Principles is given below.

SD21 – Rio Principles Scorecard

1   Human beings are at the centre of concerns for sustainable development. They are entitled to a healthy and productive life in harmony with nature.

2   States have, in accordance with the Charter of the United Nations and the principles of international law, the sovereign right to exploit their own resources pursuant to their own environmental and developmental policies, and the responsibility to ensure that activities within their jurisdiction or control do not cause damage to the environment of other States or of areas beyond the limits of national jurisdiction.

3   The right to development must be fulfilled so as to equitably meet developmental and environmental needs of present and future generations.

4   In order to achieve sustainable development, environmental protection shall constitute an integral part of the development process and cannot be considered in isolation from it.

5   All States and all people shall cooperate in the essential task of eradicating poverty as an indispensable requirement for sustainable development, in order to decrease the disparities in standards of living and better meet the needs of the majority of the people of the world.

6   The special situation and needs of developing countries, particularly the least developed and those most environmentally vulnerable, shall be given special priority. International actions in the field of environment and development should also address the interests and needs of all countries.

7   States shall cooperate in a spirit of global partnership to conserve, protect and restore the health and integrity of the Earth's ecosystem. In view of the different contributions to global environmental degradation, States have common but differentiated responsibilities. The developed countries acknowledge the responsibility that they bear in the international pursuit to sustainable development in view of the pressures their societies place on the global environment and of the technologies and financial resources they command.

8   To achieve sustainable development and a higher quality of life for all people, States should reduce and eliminate unsustainable patterns of production and consumption and promote appropriate demographic policies.

9   States should cooperate to strengthen endogenous capacity-building for sustainable development by improving scientific understanding through exchanges of scientific and technological knowledge, and by enhancing the development, adaptation, diffusion and transfer of technologies, including new and innovative technologies.

10   Environmental issues are best handled with participation of all concerned citizens, at the relevant level. At the national level, each individual shall have appropriate access to information concerning the environment that is held by public authorities, including information on hazardous materials and activities in their communities, and the opportunity to participate in decision-making processes. States shall facilitate and encourage public awareness and participation by making information widely available. Effective access to judicial and administrative proceedings, including redress and remedy, shall be provided.

11   States shall enact effective environmental legislation. Environmental standards, management objectives and priorities should reflect the environmental and development context to which they apply. Standards applied by some countries may be inappropriate and of unwarranted economic and social cost to other countries, in particular developing countries.

12   States should cooperate to promote a supportive and open international economic system that would lead to economic growth and sustainable development in all countries, to better address the problems of environmental degradation. Trade policy measures for environmental purposes should not constitute a means of arbitrary or unjustifiable discrimination or a disguised restriction on international trade. Unilateral actions to deal with environmental challenges outside the jurisdiction of the importing country should be avoided. Environmental measures addressing trans-boundary or global environmental problems should, as far as possible, be based on an international consensus.

13   States shall develop national law regarding liability and compensation for the victims of pollution and other environmental damage. States shall also cooperate in an expeditious and more determined manner to develop further international law regarding liability and compensation for adverse effects of environmental damage caused by activities within their jurisdiction or control to areas beyond their jurisdiction.

14   States should effectively cooperate to discourage or prevent the relocation and transfer to other States of any activities and substances that cause severe environmental degradation or are found to be harmful to human health.

15   In order to protect the environment, the precautionary approach shall be widely applied by States according to their capabilities. Where there are threats of serious or irreversible damage, lack of full scientific certainty shall not be used as a reason for postponing cost-effective measures to prevent environmental degradation.

16   National authorities should endeavour to promote the internalization of environmental costs and the use of economic instruments, taking into account the approach that the polluter should, in principle, bear the cost of pollution, with due regard to the public interest and without distorting international trade and investment.

17   Environmental impact assessment, as a national instrument, shall be undertaken for proposed activities that are likely to have a significant adverse impact on the environment and are subject to a decision of a competent national authority.

18   States shall immediately notify other States of any natural disasters or other emergencies that are likely to produce sudden harmful effects on the environment of those States. Every effort shall be made by the international community to help States so afflicted.

19   States shall provide prior and timely notification and relevant information to potentially affected States on activities that may have a significant adverse transboundary environmental effect and shall consult with those States at an early stage and in good faith.

20   Women have a vital role in environmental management and development. Their full participation is therefore essential to achieve sustainable development.

21   The creativity, ideals and courage of the youth of the world should be mobilized to forge a global partnership in order to achieve sustainable development and ensure a better future for all.

22   Indigenous people and their communities and other local communities have a vital role in environmental management and development because of their knowledge and traditional practices. States should recognize and duly support their identity, culture and interests and enable their effective participation in the achievement of sustainable development.

23   The environment and natural resources of people under oppression, domination and occupation shall be protected.

24 Warfare is inherently destructive of sustainable development. States shall therefore respect international law providing protection for the environment in times of armed conflict and cooperate in its further development, as necessary.

25 Peace, development and environmental protection are interdependent and indivisible.

26 States shall resolve all their environmental disputes peacefully and by appropriate means in accordance with the Charter of the United Nations.

27 States and people shall cooperate in good faith and in a spirit of partnership in the fulfilment of the principles embodied in this Declaration and in the further development of international law in the field of sustainable development.

Successes

As a soft law instrument, successful implementation of the Rio Declaration takes many shapes and can be loosely understood through analysing the various 'offspring' agreements or national laws that have transposed aspects of the Principles. Where such a transposition has occurred, and the principle has been applied in practice, its application has often been tested in the courts; the result of which is that some of these principles have been widely accepted as part of international jurisprudence.

The most prominent examples of this legal application are Principles 10 and 15, along with Principles 5, 7, 17 and 24, which demonstrate varying elements of successful transposition and wide adoption of the principle in laws. Principles 3 and 21 are steadily gaining momentum on implementation; and of latter years, in conjunction given their interrelation, both have seen an explosion of activity where increasingly more effort is being made to apply them in practice.

Principle 10 – access to justice, information and public participation – is the foundation of the very successful regional instrument that enshrines the principle in the Aarhus Convention, which applies to most EU member states as well as a handful of other acceding parties that elected to participate in it. The Aarhus Convention has provided valuable means by which the various elements of the Principle have been promoted through application at the national level, as well as providing a forum (the Compliance Committee) that can hear complaints where it is claimed that nation-states are not adhering to the convention. Notably cases have been brought by civil society organizations that have challenged their government's lack of implementation or compliance to the convention, which has resulted in the development of a body of case law that has strengthened the principle overall.

In addition, the elements of Principle 10 have been borne out in jurisdictions that are not parties to the Aarhus regional instrument, but have nonetheless used it as a persuasive example to underpin activities such as establishing national environmental courts or tribunals.

Principle 15 – the precautionary principle – is widely accepted as a foundation of environmental law at both the national and international levels. It has been tested in a range of courts and jurisdictions, notably the World Trade Organization (WTO) arbitral body where initially it was found in some cases that trade rules superseded the precautionary principle; however in more recent years this has not been the approach adopted by most states and the principle itself is well established in international jurisprudence and is increasingly becoming more accepted by different nation-states at the national level.

Principle 17 – environmental impact assessments (EIAs) – are widely employed as national instruments that are integral to the planning and development processes. Whilst the efficacy of these instruments has been challenged, the process by which EIAs as well as other strategic impact assessments have been transposed into national legal instruments provides an instructive framework for how soft law can be applied in a very practical way. The popularization of such tools demonstrates that where there has been the impetus to develop such a 'national instrument' (as defined in the Principle itself), the regulation to support it and the subsequent application of it in practice can develop with reasonable speed and intent.

Principle 24 – relating to the destructive nature of warfare – has been very well implemented in national and regional instruments. There are multiple examples of where the principle of 'respect[ing] international law providing protection for the environment in times of armed conflict', as the principle itself states, has been enshrined in national legislation and there are various international inter-governmental and non-governmental bodies that focus specifically on ensuring the successful application of these instruments. In practice, however, it has been difficult to quantify how, and if, the principle has been successful in achieving the overall objective.

Principle 5 – eradicating poverty and raising the standards of living for all – helped put the spotlight on the inequity that exists in the world and the wealth divide between rich and poor. Popular campaigns have shown that the relevance of Principle 5 reached much wider audiences than those involved in the multilateral processes, and the desire and intent to act captured the imagination of society on the whole. As such, the Millennium Development Goals (MDGs) were agreed with sincere intent to secure poverty eradication. Focusing on key indicators, the MDGs are a direct heir of Principle 5. In 2015 the MGDs expire and there will be a review of their successful application and whether or not the goals have been achieved.

Principle 7 – global cooperation to conserve, protect and restore the health and integrity of the Earth's ecosystem – enshrines the principle that was already gaining traction before UNCED, that of common but differentiated responsibilities. This concept has successfully filtered out into discussions in the multi- and bilateral regimes, at both international and national level and in specific areas ranging from climate change (under the UN Framework Convention on Climate Change –

UNFCCC) and the right to development more broadly. It is now seen as a mandatory element to every development discussion since UNCED and it underpins the core value of equity that is at the heart of many of the Rio Principles. Whilst there is not an instrument that enshrines the common but differentiated responsibility doctrine, and therefore not a specific regulatory framework similar to, for example, the instruments that enshrine principles 10 or 17, it has been accepted and adopted in international and national jurisprudence as a successful and potent development of the soft law principle.

Challenges

A critical dimension of the sustainable development concept is that of public participation in the decision-making as well as implementation process. This has been successfully adopted as practice in the various international framework regimes (the Convention on Biological Diversity (CBD), UNFCCC) and, as noted above, the concept has been successfully enshrined in instruments such as Aarhus and others. However, the lack of ability for many groups and stakeholders to participate in the process at national and local level remains an issue.

Additionally access to justice remains a barrier for many who seek legal redress for environmental damages or concerns. Notably, a claim was brought to the Aarhus compliance committee against the UK, arguing that the costs of bringing an environmental case in the UK was 'prohibitively' expensive, undermining one of the cornerstones of the Aarhus Convention. The compliance committee found in their favour, declaring the UK non-compliant to Aarhus, and time will now tell how the UK responds to such a declaration and whether this 'gap' will be filled.

Whilst the polluter-pays principle (Principle 16) has been transposed into a range of legal instruments in a number of jurisdictions and contexts, there remain ideological differences to its practical application which have undermined the successful implementation of the principle on the whole. Such ideological differences in some areas have led to the development of parallel systems which undermine the objective of the polluter-pays principle, such as waste treatment processes. In practice it has been identifies that pollution and waste continue to pervade our lifestyles and as a result of the principle not being successfully implemented, the level of pollution has not decreased.

The drive to eradicate poverty, stemming from Principle 5 as outlined above, successfully led to the MDGs; however the final aspect of the principle – that of 'reducing disparities in standards of living' – has been almost entirely forgotten, or left out of the development discussion. In some cases, it has almost become obscene in the manner in which the principle has been distracted from as attention became almost exclusively focused on income poverty. The MDGs reinforce this approach, as does the theme of the Rio conference on 'green economy in the context of poverty eradication'. It will be crucial to ensure that discussions about reducing the disparities in standards of living and wealth distribution are incorporated into the Rio+20 discussion and any subsequent regimes that stem from it.

The precautionary principle, whilst successfully implemented in a range of instruments and tested in case law, remains mired by ideological divergence, which is undermining the achievement of the overall objective of eliminating those actions that have the potential to cause serious and irreversible harm. Prominent examples where this tension is not resolved include the UNFCCC regime and wider climate-change discussions and approaches that are not adopting the precautionary principle regardless of the increasing body of evidence that indicates and underpins the importance of adopting such an approach. Many countries are not sticking to mitigation targets and many more are not taking targets on, which has the potential to result in significant irreversible harm. The GMO debate also suffers from divergences in ideology relating to the potential harm that could be caused, but which are as yet unknown.

In numerous fora, the concept of common but differentiated responsibility is widely integrated into discussions. However, conflicting approaches to this principle continue to undermine the overall objective. For example, in the climate-change regime the concept often negatively impacts on the negotiations and stalls progress.

Principles 3 and 21 focus on the concept of intergenerational equity. Justice for future generations has been a key element of sustainable development since the Brundtland Report. There have been a range of initiatives to bring the principle into the processes of decision-making at both the national and international levels and there is certainly a ground-swell of support for adopting such a principled approach in recent years. However on the whole the principle has not been reflected at the institutional level and has not had the governmental support that reflects the civil society and wider stakeholder appetite to bring the concept to the heart of sustainable development governance.

Principle 8 – sustainable production and consumption and the promotion appropriate demographic policies – is deemed to have been unsuccessful in achieving its intended goal. Unsustainable consumption patterns have continued to rise at a steady pace in industrialized countries. The BRIC countries (Brazil, Russia, India and China) are seeing blooming consumer classes that aspire to high per capita consumption levels, and other developing countries will follow suit in time. Population projections are estimating a 50 per cent rise in population by 2050 with demand for family planning services far outstripping supply at the national or local level. These trends are compounding each other and increasing the unsustainable impacts of human activities beyond the ability of ecosystems to recover.

Other specific difficulties identified in the review include:

- The contradiction between national sovereignty on resources, a fundamental tenet of Principle 2, and the widely understood concept of the tragedy of the commons that relates to trans-boundary pollution, climate change and biodiversity (as described in Principle 7 regarding global partnerships for ecosystem health) is starkly borne out as international multilateral regimes fail to adequately implement an approach to overcoming this contradiction.

- A potential contradiction in the set of principles lies between Principle 12 (growth and free trade as the model) and Principle 8 (addressing unsustainable consumption patterns). Over the last two decades it has become increasingly apparent that where and when trade and a rethinking of consumption patterns come up against one another, trade wins. This results in severe undermining of the practice of sustainable development. Overall, Principle 8 remains largely unaddressed, whereas a business-as-usual, growth-at-all-cost paradigm has dominated.

Conclusions

The Rio Principles are the heir to the Stockholm Principles (agreed in 1972), and both have a primary focus on environment and development. The construction of a whole set of principles was clearly intended to find a common ground between developed and developing countries. However this framework was less than adequate in providing a coherent vision of how to achieve sustainable development. In particular, the lack of guidelines to accompany the principles resulted in little cohesion for the implementation of the majority of the principles, and ultimately many principles remain aspirational soft law instruments that countries find difficult to transpose into national legislation without a framework of how to do so.

Overall, the social equity dimension is almost completely absent from the Rio Principles, and even after a decade of its absence it was only 'addressed' at the World Summit for Sustainable Development in Johannesburg rather than focusing on its integration into all of the principles; as such, one of the three pillars of sustainable development remains virtually absent from one of the highest level sustainable development documents that have been developed and agreed these past two decades.

In light of this, it will be especially important in the Rio+20 process to ensure that when any outcome documents are prepared (e.g. frameworks or roadmaps) there is provision for a set of guidelines or a guidance notes on implementation. This, with a stronger intent to deliver on agreed outcomes and a deeper political will, could go some way to ensuring actual implementation is far greater than what we have seen over the past twenty years.

The original Rio deal

Rio recognized the need to redirect international and national plans and policies to ensure that all economic decisions took into account environmental impacts. The deal arising from Rio took a three-pronged approach:

1 developed countries would take the lead in changing production and consumption patterns (their economic model);
2 developing countries would maintain their development goals but take on sustainable development methods and paths;

3   developed countries would enable and support the developing countries' sustainable development through finance, technology transfer and appropriate reforms to the global economic and financial structures or practices.

Issues requiring an integration of economic and environmental concerns (such as climate, the interaction of trade and environment; and the relation between intellectual property rights and environmental technology and indigenous knowledge) were to be resolved through international cooperation, in which the development needs of the developing countries would be adequately recognized.

What happened?

Despite this well-meaning deal, reality has fallen considerable short of ambition. Significantly developed countries did not curb their consumption patterns and failed to find a sustainable development path built on sustainable production methods. As a result, pressure on the global environment continued to rise since 1992.

Specifically, despite continued intergovernmental process (e.g. climate-change talks and further Earth Summits), little progress has been made toward implementation of the deal. Most recently an international agreement on climate change has all but stalled. Funding arrangements and transfers of technology from developed to developing nations around the Agenda 21 outcomes have not been delivered as promised. No 'additional resources' were provided to facilitate the transition. In fact, overseas development aid (ODA) fell from $US62.4 billion in 1992 to $US48.7 billion in 1997. It was not until 2002 that it again topped the $US60 billion mark. This 'lost decade' was marked by regression of key development statistics with many of the world's poorest countries suffering from worsening poverty. However, aid flows from donor countries totalled $US129 billion in 2010, the highest level ever, and an increase of 6.5per cent over 2009 (OECD, n.d.).

At the Monterrey Financing for Development Conference in 2002, world leaders pledged 'to make concrete efforts towards the target of 0.7per cent' of their national income in international aid. However, as of 2003, only five countries have already met or surpassed the 0.7per cent target: Denmark, Luxembourg, Netherlands, Norway and Sweden (the UK may become the sixth in 2013). In 2005, total aid from the 22 richest countries to the world's developing countries was just $US106 billion – a shortfall of $US119 billion dollars from the 0.7per cent promise (Sachs, 2005). ODA flows are slowly being supplemented by funding sources (e.g. foreign direct investment (FDI), remittances and future global private equity fund schemes). However, these are nowhere near the magnitude of ODA, and these flows are not factored in the thinking of what ODA can deliver.

At the same time, there came a realization that the implicit basis for the compromise, which was that globalization in the form of economic growth plus free trade would lift all boats, was not delivering automatic dividends and was in fact further marginalizing some developing regions. Developing nations felt that they were short-changed on trade issues, and in fact, this global economic paradigm served to increase inequality and has exacerbated environmental issues, which have further worsened human well-being in many regions. Due to the lack of change delivered by the historical development model, the major developing countries are following the developed countries' model of development and the pressure on the planet is increasing.

In 2000 the Millennium Development Goals (MDGs) were established following the Millennium Summit. The aim of the MDGs was to encourage development by improving social and economic conditions in the world's poorest countries. However, in the last decade, the MDGs have taken the focus off the larger sustainable development agenda and focused on a much narrower set of goals – which did not address any of the fundamental drivers, even more so than Agenda 21. The MDGs were adopted as *the* reference framework by the development community, leading to the aim of alleviating poverty without properly addressing underlying causes. For example, the MDGs were focused solely on developing countries and did not address consumption issues of developed countries which have clear knock-on effects in developing countries. Also, resources started to flow to climate change-related issues, further marginalizing sustainability as the integrated concepts needed to resolve interconnected issues.

During the 2000s, a divergence of outcomes developed among former developing countries. Despite still officially speaking with a unified voice there was recognition of the divergent interests and needs between countries. It has become increasingly clear that a more nuanced approach to development is needed which takes into account local culture, environmental and social pressures, and regional ecosystem services. However, in practice, ODA is often unpredictable, untargeted and does not make it to where it is needed. It is estimated that 'only about 24% of bilateral aid actually finances investments on the ground' (Sachs, 2005). Therefore it is clear that better aid is needed, not just more aid. Improving the quality of aid and ensuring it is delivered on the ground in line with local requirement is as important to sustainable development as is increasing the amount of aid.

The move towards adopting Sustainable Development Goals, an initiative by Colombia and Guatemala in 2011, could go a long way to creating a more substantive and integrated framework to bringing together the agendas of Rio and the MDGs and ensuring there is only one development model – a sustainable development one.

Resolving contradictions

The international developments on sustainable development have given rise to a contradiction. On the one hand, the Brundtland Commission put to the fore two critical dimensions in its report and definition on sustainable development: 1) caring about future generations and 2) addressing unsustainable consumption patterns of the rich (Rio Principle 8).

On the other hand, one way of seeing the Rio outcomes (Rio principles + Agenda 21) is as a 'business as usual plus (BAU+)' arrangement. The fundamental assumptions of post-war neo-liberal economics (i.e. economic growth coupled with free trade) were left unchallenged. Instead they were simply adorned with 'safeguards' that satisfied both North and South. This has resulted in environmental safeguards for the North which ensured discreet environmental issues were managed, and development safeguards for the South which ensured economic development could continue unimpeded.

Implicit was a hope that BAU+ was able to deliver sustainable consumption and production patterns and longer-term decision-making, and that these were compatible in practice through decoupling of resource use from consumption. However, there was quickly no doubt left about which of the two would prevail when conflict arose or decoupling proved to be difficult to achieve. Business as usual has prevailed and unsustainable patterns of consumption and production persist. This in turn means that global commons (e.g. forests, atmosphere, biodiversity, oceans) are still managed unsustainably and, worse, are degraded beyond their ability to recover unless pressure is lessened.

It has now become clear that increases in humanity's environmental impact are rising, and in the future this will be compounded by rising population and increased demand for affluence. It is unlikely that these trends will be countered by improvement delivered through technology as dictated by the IPAT model (i.e. human impact (I) on the environment equals the product of P = population, A = affluence and T = technology). Therefore acting on the 'technology' factor alone cannot do the job, based on past experience (Jackson, 2009). So, change is needed. A corresponding change in behaviour is urgently required to balance the equation and ensure the anthropomorphic environmental damage is kept in check, and that development patterns deliver an equitable future within environmental limits.

This will require that other contradictions are also urgently addressed as they stall progress or obscure issues. These include:

- sovereignty versus global goods – ensuring that the sovereign right to exploit resources (Rio Principle 2) does not override the global partnership to conserve, protect and restore the health and integrity of the Earth's ecosystem (Rio Principle 7).
- precaution versus markets – adopting a precautionary approach to ensure that a lack of full scientific certainty shall not be used as a reason for postponing cost-effective measures to prevent environmental degradation (Rio Principle 15).
- polluter pays – ensuring that the polluter should bear the cost of pollution, with due regard to the public interest and without distorting international trade and investment (Rio Principle 16).

It should be clear that the 'market' outcomes have to be more regulated based on principles that put values first (e.g. the fundamental principles enunciated in the Millennium Declaration: freedom, equality, solidarity, tolerance, respect of nature and shared responsibility). Currently there is a lack on linkage between commonly agreed values and principle and market practices. This has clear links to the underdeveloped social dimension within the Rio Principles. Further development of such values will help shift behaviours, drive practices and ultimately achieve sustainable outcomes (such as those described by the Rio Principles).

These challenges have been clearly written about over the last ten years. Ashok Khosla put it plainly in 2002 when he said:

> But first, let us look at the current gap that most deeply threatens the whole process of international negotiation to which we are all so strongly committed. This is, of course, the Implementation Gap. Closely related to it is the Accountability Gap. Who is asking what it is that we are doing about the promises we made at Stockholm, at Nairobi, at Mexico, at Dublin, at Tblisi, at Paris, at Rome, at Rio, at Istanbul, at Beijing, at Copenhagen, at Cairo or at any of those other wonderful places that our diplomats love to travel to? This is a gap that is widening precipitously. Every time we hold a conference we promise more, and we do less. And there is no one to hold us accountable, not even civil society. Governments took on the responsibility but they are too busy scratching each other's backs.
>
> (Khosla, 2002)

We have to do things differently.

Chapter 11

The democracy gap

Democracy is not only a political system … It is an ideal, an aspiration, really, intimately connected to and dependent upon a picture of what it is to be human – of what it is a human should be to be fully human.

(Kompridis, 2009)

A stakeholders' democracy

As has been highlighted in earlier chapters, the role of stakeholders at the United Nations has become far more significant over the last 40 years, and particularly in the 20 years since the Earth Summit.

This has occurred at the same time as the number of democracies in national governments has been increasing – from 40 in 1972 to 115 in 2010 (a figure that doesn't reflect the results of the Arab spring, but which is down from its high of 123 in 2005, according to the *Freedom in the World 2011* report (Puddington, 2011). The opening up of previously closed societies has been contributed to by the participation from stakeholders in those countries, and has in turn encouraged the further growth of such participation at all levels.

This changing geo-political landscape has coincided with many of the more positive impacts of globalization, bringing with it new information technologies, low-cost forms of communication, and an inevitable interconnection of global affairs. An obvious by-product of this development has been the rapid growth of an increasingly integrated global civil society – a phenomenon rarely seen before 1992, which has had a profound impact on the mechanisms of global governance and democratic practices. This point was already being emphasized in 1999 by Kofi Annan, UN Secretary-General at the time:

The UN once dealt only with governments. But now we know that peace and prosperity cannot be achieved without partnerships involving governments, international organizations, the business community and civil society. In today's world, we depend on each other.

(Annan, 1999)

The role that stakeholders could play might not even have been fully anticipated in 1992 when the idea of the independent sector first was reflected in a UN document at the Earth Summit, which framed it as the Major Groups.

Since then there has been a plethora of experiments in stakeholder participation at the local, subnational, national and global levels of government – and also an increasing degree of stakeholder involvement with enlightened businesses. This suggests a need now to reflect and develop a coherent set of approaches to stakeholder democracy at all levels.

The emergence of a stakeholder space

Civil society is now so vital to the United Nations that engaging with it well is a necessity, not an option.

(Cardoso, 2004)

The last 20 years have shown that we are in the middle of what has been called a transition from Madisonian (or representative) democracy to Jeffersonian (or participatory) democracy, and are now entering in a period of democracy by, of, and with stakeholders. Stakeholder democracy is the idea that involving all participants, at every level, will result in better-informed decisions being taken. It means that those stakeholders will thereby feel more ownership of the outcome and then be active in the delivery of policy on the ground, often in partnership with governments and other stakeholders.

At a time of increasing disillusionment with government, stakeholder democracy offers a platform that could help give governments the moral and political strength to address the difficult decisions that need to be made now and in the near future to ensure the viability, equitability and sustainability of life on this planet.

The 'occupy' movement for greater economic equality, which is growing as we write, is an expression of how dissatisfied a large portion of the general public has become with the extraordinary mess that a majority of the developed world's economies have fallen into. Over the last 20 years, as governments have deregulated the financial sector, the rich have become obscenely richer and the poor tragically poorer. The global banking crisis that exacerbated these polarities is now perversely blocking action to reform the very failures that brought them about. It is yet unclear how politicians will react to the expression of popular outrage represented by the new movement, or whether it will solidify into a coherent force with more clear political demands. But it is clear that governments can and must act.

Governments and stakeholders at Rio in 2012 must agree that the model that the world tries to move towards is not just a 'green economy or economies' or a 'lean economy', but also one that is not an 'obscene economy' featuring gross disparities of opportunity and wealth, but rather is based on equity and fairness for all. With the real possibility of major crises in the near future in such areas as food, energy and water – and with the now unavoidable increased impacts of climate change –

there urgently need to be more, and more stable, national and global economies, and better ways of addressing problems of economic governance.

Stakeholders are not limited by national boundaries. A single organization can be active in many countries either through international membership, parallel autonomous national chapters, or working in multinational coalitions with other organizations. To some extent they can bring together global and local perspective in a way governments, and even intergovernmental agencies, cannot. And where better to express that than within the United Nations processes? So when the UN opened its doors to participation by NGOs and others, it found a highly complementary fit. The UN provides a space to address the present and future challenges all parties face. Governments have political, economic and military power, but are all too often restricted by perceptions of national interests or sovereignty. Stakeholders lack discrete sovereign power, but can integrate ideas, present advice on policy, and rally political backing.

Of course the involvement of stakeholders isn't entirely new. The International Labour Organization (ILO), established in 1919, has a tripartite governing body composed of 28 government representatives, 14 workers' representatives, and 14 employers' representatives. It has functioned effectively and has even been looked at as a model for changing global governance. However, the ILO model was never replicated on the international level.

Major non-governmental international organizations such as IUCN have served in a respected advisory capacity at UN meetings since its founding. And traditional humanitarian NGOs, like Oxfam and CARE, have for decades been contracted to provide services on the ground. What is new, however, is the active integration of all these roles – advising, lobbying, implementing and building of constituencies – by a broad and diverse population of organizations representing virtually every country, every issue area and every demographic sector.

Reforming the UN for stakeholders

> Over the last twenty years we have seen a dramatic increase in the number of civil society organizations (CSOs) taking an active role in UN intergovernmental meetings. This participation has ranged from interactive dialogue sessions with ministers, to civil society hearings, to invitations for CSOs to take part in roundtable discussions, panel discussions or parallel meetings. Such initiatives have been critical for developing processes that can underpin the inclusion of non-governmental actors in international meetings, and have gone some way to ensuring more accountable, transparent and informed global policy making.
> (Benson, 2010)

The two critical documents that helped codify the international empowerment of stakeholders were each created at the Earth Summit. One was the Rio Declaration's Principle 10, on access to information, publication participation and environmental justice. The second, of course, is Agenda 21 and its nine chapters on Major Groups.

The decade of intergovernmental summits and convention meetings in the 1990s included a series of interesting experiments. The Habitat II Conference, in 1996, for the first time at a UN conference provided non-voting seats at the negotiating table for stakeholders. This approach – started in the informal meetings prior to the conference – allowed stakeholders to put forward amendments to the negotiating text that became part of the draft text if any government endorsed the amendment. Habitat, also for the first time, then published those stakeholder amendments as an official UN information document – whether they were accepted by governments or not.

At preparatory meetings of the World Summit on the Information Society, a Civil Society Bureau and a Private Sector Bureau were both established. The bureaus met together at times throughout the process to deal mostly with organizational issues.

The Joint United Nations Programme on HIV/Aids (UNAIDS) Board established in 1996 has five representatives, and five alternates represent the perspectives of civil society; these are non-voting members. UNAIDS was the first United Nations programme to have formal civil society representation on its governing body.

It is not surprising therefore that the Aarhus Convention – a convention dealing with public participation agreed by the UN Economic Commission for Europe, which flowed directly from the Rio Declaration Principle 10 – allows NGOs the status of 'non-voting participants' during negotiations. It has evolved to let NGOs basically participate in negotiating processes. An environmental representative can also sit on the Aarhus Convention's coordinating bureau.

A final example is in the Strategic Approach to International Chemicals Management (SAICM), a process set up after the WSSD, which also involves stakeholders sitting in its governing body and in negotiations as equal partners.

Looking at these different examples, Stakeholder Forum – working with an advisory board composed of UN 'focal points' for Major Groups –suggested a 'Stakeholder Standard' to help the UN system as a whole to learn from the experience of the last twenty years.

The Agenda 21 chapters on Major Groups were very influential in the first 10 years after their approval in gaining space at all levels of governance. Now, with nearly 20 years' experience, there's a need to review the concept and consider some potentially useful reforms.

Why only the nine?

As defined by Agenda 21, the nine Major Groups are an epistemologically and linguistically incoherent set. They consist of three groups defined by their economic function (business, trade unions and farmers), three defined by their demographic sector (indigenous people, women and youth) and three defined broadly by their institutional role in society (NGOs, local authorities and the scientific community).

The UN CSD was the first to use the Major Groups structure at the global level, as it was the primary body reviewing Agenda 21 implementation. It has been followed in recent years by other environment-focused bodies such as UNEP

Box 11.1 **The Stakeholder Standard**

1. Allow at least six months for stakeholder coordination and consultation prior to an intergovernmental meeting.
2. Clearly define 'civil society' into groups, constituencies or stakeholders who will be affected by the outcome of the intergovernmental meeting.
3. Ensure all civil society coordinators represent a wider network or stakeholder group and have the resources and mechanisms to reach out to those networks.
4. Involve a combination of stakeholders in the initial design and preparatory processes for an intergovernmental meeting.
5. Identify and publish a set of criteria for the appointment of key civil society partners and coordinators. Involve stakeholders to define those selection criteria.
6. Ensure that summaries of all meetings with key civil society partners and coordinators are made publicly available.
7. Cover the costs for travel and accommodation for key civil society partners and coordinators.
8. Include capacity-building and training initiatives as part of the civil society engagement strategy.
9. Provide all civil society coordinators with the same information and the same documents at the same time.
10. Allow time and resources for the engagement strategy to be evaluated and assessed after the meeting has come to an end.

Source: Benson, 2010

---

and the UNFCCC. The Convention on Biological Diversity has interpreted *its* Major Groups to include some similar and some curiously different categories of stakeholders (local authorities, business, parliamentarians, universities, the science communities, children and youth, NGOs, and the Green Wave for schools).

The nine stakeholder sectors selected in 1992 are not the only stakeholders that could or should be involved in the process. Among the obvious 'missing sectors' that have been suggested are the faith community, the education community, subnational governments, parliamentarians, regional government and the (soon to be numerically exploding) demographic of older people. But amending the process to add such categories now would face typical political inertia, plus the real obstacle of the sheer number of sectors becoming logistically too unwieldy to provide adequate time for each.

The decision on the nine wasn't taken with an extensive discussion of the possible implications in the future. Rather, it was a more mundane exercise in pragmatic geopolitical political give and take. The first group included was women, because at the summit the international caucus of women's environmental organizations had played an early and highly effective role in advocating issues and mobilizing public support.

UN organizations have tended to accept the nine because they have an intergovernmental mandate in Agenda 21, although the original mandate provided to the CSD by the General Assembly doesn't actually mention the term 'Major Groups'. It simply recommends:

> … that the Commission have the following functions, as agreed in paragraphs 38.13, 33.13 and 33.21 of Agenda 21:
>
> (f) To receive and analyse relevant input from competent non-governmental organizations, including the scientific and the private sector, in the context of the overall implementation of Agenda 21;
>
> (g) To enhance the dialogue, within the framework of the United Nations, with non-governmental organizations and the independent sector, as well as other entities outside the United Nations system.
>
> (United Nations, 1992b)

The WSSD in 2002 affirmed the value of stakeholders in general, especially in achieving partnerships and implementation. There was strong lobbying of governments in 2002 by a number of key stakeholders, such as the education community and regional governments, for the UN secretariat to take a more relaxed approach as to which stakeholders could be involved, and specifically to add a Major Group to cover 'educators'. In the end, the Johannesburg text included only the following reference:

> 146 Within that context, the Commission should place more emphasis on actions that enable implementation at all levels, including promoting and facilitating partnerships involving Governments, international organizations and relevant stakeholders for the implementation of Agenda 21 …
>
> 149(b) Continuing to provide for more direct and substantive involvement of international organizations and major groups in the work of the Commission.
>
> (United Nations, 2002b)

It did at least seem to support the role of educators.

> 149(d) Furthering the contribution of educators to sustainable development, including, where appropriate, in the activities of the Commission.
>
> (United Nations, 2002b)

In the post WSSD period the nine Major Groups gained formalization within the UN CSD, and then within broader parts of the UN system. This allowed those groups to stop fighting for recognition, though it ironically may have made it more difficult for other voices to be heard.

There was experimentation around the civil society structure at the Bonn Energy Conference in 2004, a key intergovernmental meeting, though not directly related to the UN. Here CBOs and NGOs were represented separately, and the industry group was split into three – traditional energy companies, renewable companies and the finance sector. It was a model that the UN system might find useful for a process such as climate change or energy.

Coordinators? Facilitators? Gatekeepers?

Major Groups are not homogeneous. The different views of a multitude of NGOs or women's groups or youth or indigenous peoples are often difficult to integrate into one voice. That is also true of the business sector – particularly in critical negotiations such as climate change where there can be significantly differing views among traditional energy companies, renewable energy companies and the insurance/finance sector. So the internal process by which each of those sectors of stakeholders coordinates among its members becomes important.

It is here that the challenge of acting on democratic principles becomes interesting. How do you coordinate decision-making in a coalition that includes dozens, or even hundreds, of discrete organizations? Do you put every question up for a vote? Do you set up a system of 'one organization, one vote', or of 'one vote for every attending individual representative'? Do you give equal weight to votes from an organization of 50,000 members with a staff of 300, and to an organization of 50 members with a staff of three? Do you require absolute consensus agreements, or go for majority rule – or for a 60 per cent minimum? How do you deal with organizations or individuals whose sole purpose seems to be slowing down agreements and obstructing actions? The answers are not easy, and there are no absolutes.

Discussing and reaching decisions in a coalition of hundreds of organizations can be painstaking and time-consuming. Each of the many highly devoted individuals involved often has his or her own distinct and carefully formulated opinion of the way things should be. The search for total consensus – a philosophical ideal of many NGOs – can result in a 'lowest common denominator' decision to ensure that everyone is on board.

Organizations with 50,000 members and professional staff can become impatient waiting for agreement by organizations staffed by three volunteers. Often such 'major' NGOs simply decide to, as the time spent in negotiating common positions takes them away from lobbying.

Similarly, the process by which each stakeholder sector coordinates with the other sectors – and with the governmental or intergovernmental process – carries great significance. Does each stakeholder sector have its own relationship to a conference's secretariat, or is there a common facilitator on the official side? Are all the stakeholder sectors tasked to consult together? Is each sector always equally included in every debate, or are certain sectors given higher profiles when their interests are more relevant to a specific issue?

In these and many other delicate areas the individuals in the coordinating leadership roles have to constantly tread a fine line between pushing enough to be effective in obtaining agreements for action, and pushing too much and risking becoming restrictive 'gatekeepers' accused of limiting access to full participation.

Multi-stakeholder platforms

Over the last ten years a series of additional multi-stakeholder platforms have emerged, organized along Major Group lines, which have significant potential for success but no obvious way to interface with the UN system. Processes such as the Sustainable Agriculture and Rural Development (SARD), the Water and Climate Coalition (WACC) and the Green Economy Coalition (GEC) require consideration of whether the whole approach to engaging stakeholders needs to become more flexible. As Paul Hohnen, the former Strategic Director of Greenpeace International, described the new multi-stakeholder space: 'Multi-stakeholder processes (MSPs) are a new species in the political ecosystem. They will make mistakes. They will not solve all problems to everyone's satisfaction' (Hohnen, 2002). These experiments have also revealed certain key principles that can guide relationships within that space. A set of these have been described in Box 11.2.

Box 11.2 **Key principles and strategies of stakeholder democracy**

Accountability: Employing agreed, transparent, democratic mechanisms of engagement, position-finding, decision-making, implementation, monitoring, evaluation; making these mechanisms transparent to non-participating stakeholders and the general public.

Effectiveness: Providing a tool for addressing urgent sustainability issues; promoting better decisions by means of wider input; generating recommendations that have broad support; creating commitment through participants identifying with the outcome and thus increasing the likelihood of successful implementation.

Equity: Levelling the playing field between all relevant stakeholder groups by creating dialogue (and consensus-building) based on equally valued contributions from all; providing support for meaningful participation; applying principles of gender, regional, ethnic and other balance; providing equitable access to information.

Flexibility: Covering a wide spectrum of structures and levels of engagement, depending on issues, participants, linkage into decision-making, time-frame,

---

and so on; remaining flexible over time while agreed issues and agenda provide for foreseeable engagement.

Good governance: Further developing the role of stakeholder participation and collaboration in (inter)governmental systems as supplementary and complementary *vis-à-vis* the roles and responsibilities of governments, based on clear norms and standards; providing space for stakeholders to act independently where appropriate.

Inclusiveness: Providing for all views to be represented, thus increasing the legitimacy and credibility of a participatory process.

Learning: Requiring participants to learn from each other; taking a learning approach throughout the process and its design.

Legitimacy: Requiring democratic, transparent, accountable, equitable processes in their design; requiring participants to adhere to those principles.

Ownership: People-centred processes of meaningful participation, allowing ownership for decisions and thus increasing the chances of successful implementation.

Participation and engagement: Bringing together the principal actors; supporting and challenging all stakeholders to be actively engaged.

Partnership cooperative management: Developing partnerships and strengthening the networks between stakeholders; addressing conflict; integrating diverse views; creating mutual benefits (win–win rather than win–lose situations); developing shared power and responsibilities; creating feedback loops between local, national or international levels and into decision-making.

Societal gains: Creating trust through honouring each participant as contributing a necessary component of the bigger picture; helping participants to overcome stereotypical perceptions and prejudice.

Strengthening of intergovernmental institutions: Developing advanced mechanisms of transparent, equitable, and legitimate stakeholder participation strengthens institutions in terms of democratic governance and increased ability to address global challenges.

Transparency: Bringing all relevant stakeholders together in one forum and within an agreed process; publicizing activities in an understandable manner to non-participating stakeholders and the general public.

Voices, not votes: Making voices of various stakeholders effectively heard, without dis-empowering democratically elected bodies.

Source: Hemmati et al., 2002

The combination of the Stakeholder Standard and these principles provides a basis for stakeholders to now suggest a more formal role with the UN institutions they are engaging. The possibility of new institutions in the area of environment and sustainable development offers the chance to integrate those new ideas into any new governance structure.

National councils/commissions on sustainable development

The United Nations Commission on Environment and Development (1987) suggested that countries may 'consider the designation of a national council, public representative or 'ombudsman' to represent the interests and rights of present and future generations' (WCED, 1987). The Earth Summit underlined this in chapters 8 and 38 of Agenda 21, which recommend the establishment of national councils on sustainable development (NCSDs) to ensure the development and implementation of strategies and policies on sustainable development. By the WSSD in 2002, there were over 100 national councils (or equivalent) around the world. These were established through widely diverse mechanisms, such as:

- Presidential decree: Argentina and Vietnam.
- Ministerial decree: Niger and Barbados.
- Council of State decision: Finland.
- National law: Mexico and Philippines.
- Letter from the environment minister: Norway.
- Cabinet resolution: Ukraine and Grenada.

The structures of NCSD differed from country to country but had some common attributes such as:

- consensus-building;
- engagement and partnership;
- fair process; and
- transparency.

(IIED, 2005)

Over the years these NCSDs have achieved a number of successes. They have proven to be a very effective mechanism for governments to consult with stakeholders and independent sectors; in doing so they have helped to build support for potentially difficult legislation. They have also produced national policies and strategies on sustainable development that have advanced parts of Agenda 21 and the WSSD JPoI.

There have also been weaknesses in the enabling conditions of NCSDs. They are easily abolished if not created under a legal mandate as governments or priorities change. They need to be adequately funded, and most have not been. The most successful ones have been linked to the office of the prime minister (e.g. Finland and the Philippines), and that usually has not been their structure. Even when supported by strong political mandate, the NCSDs have not necessarily survived. At present there are 14 NCSDs in the European Union, and the European Strategy on Sustainable Development (European Union, 2006) declares that:

> Member States should consider strengthening or where they do not yet exist, setting up multi-stakeholder national advisory councils on sustainable development to stimulate informed debate in the preparation of National Sustainable Development Strategies and/or contribute to national and EU progress reviews. National sustainable development councils are meant to increase the involvement of civil society in sustainable development matters and contributes to better linking different policies and policy levels.

Yet the National Sustainable Development Commission (NSDC) of the UK was abolished in 'cost-reducing' moves by the centre-right government in 2010. The move away from national multi-stakeholder bodies is not a good sign.

One of the outcomes from the United Nations Conference on Sustainable Development 2012 will be to give an opportunity to reactivate and reinvigorate the national multi-stakeholder fora with mandates for:

- outreach, for consultation and feedback, to their constituencies in the country;
- development of guidance on implementation strategies within a country for moving to economies that put sustainable development at the heart;
- review of development of national reports on implementation; and
- development of national targets for policy, strategies and future implementation.

In light of the focus of UNCSD 2012 on the economy and how to make it compatible with sustainable development, such councils might consider establishing or strengthening their relationship with national economic councils which already exist in many countries. The reports of such councils could be integrated within parliamentary procedures reporting to both government – heads of state – and the relevant parliamentary oversight committee.

The subnational level

> We believe that the implementation of sustainable development needs a strategic framework for all governments. We believe that this applies strongly in our regional spheres. Regional governments need sustainable development strategies as central frameworks for linking all their other strategies, ensuring that each is sustainable and that they are mutually supportive of each other.
>
> (NRG4SD, 2002)

Parallel to WSSD, a new body was created to deal with sustainable development at the subnational levels, that is, levels of government immediately below the national level, such as major provinces and cities. It was called the Network of Regional Government for Sustainable Development.

Subnational governments, because of their intermediate position between the national and local levels, are particularly well placed for identifying the needs and capabilities of their societies for sustainable development. They are also often responsible for the elaboration and implementation of policy, legislation programmes, fiscal mechanisms and public investment plans in areas such as climate action, transport, energy, the environment, agriculture, forestry, industry, spatial planning, resource management, technology development and transfer, civil protection, and development cooperation – all of which directly influence sustainable development goals.

Subnational governments have been effective at addressing sustainable development challenges, as well as the economic difficulties many of them are currently facing, by turning these challenges into opportunities to transition towards greener, smarter and more inclusive societies.

Since WSSD, subnational governments have implemented extensive initiatives that have facilitated the success of multiple sustainable development policies (such as Agenda 21, climate-change adaptation and mitigation strategies, nature and biodiversity conservation measures, public procurement strategies and development cooperation projects). Recently the network has identified a series of significant challenges that must be met to accelerate the pace towards global sustainability:

- Incorporation of an explicit territorial dimension into the implementation tools and mechanisms of the different multilateral sustainable development agreements, coupled with sufficient resources for multilevel implementation.
- Further encouragement of, and adequate support to, the elaboration of interministerial and overarching sustainable development strategies at national and subnational level.
- Independent, cross-ministerial sustainable development commissions in each country, based on multi-stakeholder and multilevel governance membership.
- Development and implementation of standard sustainability assessment methodologies for all levels of government based on, for example, standard indicators and objectives, standard calculations of externalities. Further use of sustainable and green public procurement at all levels of government.

Subnational governments should and must in the future play a much more significant role. The establishment multi-stakeholder subnational councils on sustainable development and the green economy would be an important development. Such councils would report to the regional head of government and should again be integrated to report to the relevant subnational government committee.

Local stakeholder democracies: agree globally, implement locally

The emergence of what became known as Local Agenda 21 out of the Earth Summit was frequently a significant factor in framing local debate on sustainability in the 1990s and on subsequent policy. Agenda 21 recognized that many of the problems and the solutions in addressing sustainable development would have their roots at the local level. Chapter 28 declared that 'By 1996, most local authorities in each country should have undertaken a consultative process with their populations and achieved a consensus on "a local Agenda 21" for the community' (United Nations, 1992a).

By 2001, over 6,000 local authorities in 113 countries had started a dialogue with their citizens based on that declaration. It was one of the first times a UN document had such a profound impact on engaging activities at that level. Prior to this, most UN outcomes simply focused on the actions of national governments. It had been the general understanding that those governments should follow through on UN agreements, and had the will and ability to do so.

By 1992 that view was starting to wear thin. It was clear that because local authorities were the closest level of government to the people, they could play a significant role in educating and promoting the kind of implementation that sustainable development demands. One of the lead actors in encouraging such local coordination is the Local Governments for Sustainability coalition, originally founded as the International Council for Local Environmental Initiatives (ICLEI). With 1200 municipal and regional members, it has played a significant role in helping councils to share experiences, develop plans and input policy at the UN level.

Following up the Earth Summit, representatives of European cities and towns met in 1994 and agreed a set of sustainability principles in the Aalborg Charter. This was followed by the Lisbon Action Plan (From Charter to Action, 1996), the Hannover Call of European Municipal Leaders at the Turn of the 21st Century (2000), and the Johannesburg Call (2002). Then, in 2004, European local authorities met again in Aalborg and agreed the Aalborg Commitments. The commitments consolidated ten years of local sustainability efforts into more specific commitments under nine broad themes (Box 11.3).

After 2004, many local councils focused more specifically on targets and action on climate change, and then on biodiversity issues, as those agendas gained global prominence.

These experiments in local democracy have faced some detractors, but in fact those are based on the substance of the issues under consideration, not on the principles of local democracy. In the United States very recently the right-wing and faux-populist Tea Party has attacked Local Agenda 21 in the same absolutist, isolationist terms that it has reacted to all efforts to achieve progress toward international cooperation or sustainability. A standardized speech used across the US includes:

> Agenda 21 and its clandestine component Sustainability are a direct assault on our private property rights and American sovereignty and it is being implemented here in Las Cruces. Our City Council is on a path to re-engineer our society and destroy our property rights.
>
> (Tea Party, 2011)

Box 11.3 **The Aalborg Commitments**

Governance: We are committed to energizing our decision-making processes through increased participatory democracy.

Local management towards sustainability: We are committed to effective management cycles, from formulation through implementation to evaluation.

Natural common goods: We are committed to fully assuming our responsibility to protect, to preserve and to ensure equitable access to natural common goods.

Responsible consumption and lifestyle choices: We are committed to adopting and facilitating the prudent and efficient use of resources and to encouraging sustainable consumption and production.

Planning and design: We are committed to a strategic role for urban planning and design in addressing environmental, social, economic, health and cultural issues for the benefit of all.

Better mobility, less traffic: We recognize the interdependence of transport, health and environment, and are committed to strongly promoting sustainable mobility choices.

Local action for health: We are committed to protecting and promoting the health and well-being of our citizens.

Vibrant and sustainable local economy: We are committed to securing inclusive and supportive communities.

Social equity and justice: We are committed to creating and ensuring a vibrant local economy that gives access to employment without damaging the environment.

Local to global: We are committed to assuming our global responsibility for peace, justice, equity, sustainable development and climate protection.

The information being put out is often inaccurate and inflammatory. The speech goes on to say about Agenda 21: 'It is the goals of this insidious plan that are undermining and destroying our way of life.' A recent campaign attempts to push US local governments to stop promoting sustainable development and to leave ICLEI. Curiously, the same groups that frequently criticize national government for addressing issues that they argue should be decided locally, also attack local governments for attempting to implement the same policies.

The challenge for the next period will be to see local governments assume an even more important role in making local economies more sustainable. The creation of green jobs will in many cases mean local jobs, and through strategies like local green bonds local authorities can support this. The next 20 years should see local authorities and their communities developing planning that allows construction only of green buildings and the refurbishment of old buildings to new green standards, that builds a transport system using less carbon-based fuels, that improves the quality of life of its citizens, and that by doing all of these this creates thousands of local jobs. To enable this, local authorities should establish sustainability councils to guide and support their localities' planning and transitions.

At the global level, local authorities should, along with other stakeholders have a seat at the table in the global institutions responsible for environment and sustainable development.

Occupying the future

The argument that stakeholders basically use a platform to be 'against something' (i.e. to oppose a particular government or a policy or an institution) is now not so easy to make. Instead, we are seeing the evolution of an agreed space where various stakeholders try to cooperate with governments and institutions to make better decisions, and then participate themselves in the implementation. There will still be occasions for stakeholders to talk against governments or particular policies which do wrong or fail to fulfil commitments But there is now increasingly a place where stakeholders take a proactive role, and governments play the role of 'facilitator' (to use a favoured term of former US President Bill Clinton).

A stakeholder-based model of democracy provides huge potential for addressing global challenges and mobilizing local resources. And at a time of rapid political change and increasing popular alienation, it provides mechanisms for stabilizing democracies. But despite the increasing acceptance of democracy on the national level, and the recognition of a multitude of stakeholders on the international scene, in an uncertain world, participatory, stakeholder democracy will require continued, steady advocacy and implementation to move from newly established principle to accepted political reality.

These changes demand an opening up, not a closing down, of our societies. The creation and maintenance of healthy democracies requires ever vigilant and vibrant stakeholder participation. And there may be more than one model that's effective.

Adopting multi-stakeholder processes does not mean that everything will become easy. But it is clear that the present representative democracy model is encountering serious obstacles. Stakeholder democracy enables a way for more voices to be heard, and for more individuals motivated by cooperation to act. Former minister of water of South Africa, Kadar Asmal, put it well:

> A parting warning: [conducting an MSP] is never a neat, organized, tidy concerto. More often, the process becomes a messy, loosely-knit, exasperating, sprawling cacophony. Like pluralist democracy, it is the absolute worst form of consensus building, except for all the others.
>
> (Asmal, 2000)

Chapter 12

The economic gap
Transforming the economy

Economic growth as we have known it is over and done with ... The economic crisis that began in 2007–2008 was both foreseeable and inevitable and it marks a permanent, fundamental break from the past decades – a period in which most economists adopted the unrealistic view that perpetual economic growth was necessary and also possible to achieve. There are now fundamental barriers to ongoing economic expansion and the world is colliding with those barriers.

(Heinberg, 2011)

Up against the Wall Street

In the spring and summer of 2008 – precisely 40 years after the halls of consumer materialism were being challenged by straggly bands of barely organized youth, and less than 20 years after the collapse of world communism – global capitalism was itself brought to the verge of the abyss, not by outside armies or domestic rebels, but by the excesses of its own elite practitioners.

The global financial crisis has provided abundant 'teachable moments' for politicians, policy-makers and the public to ponder a series of critical lessons. But clearly, too many of them have not yet learned enough. The real questions will be, first, whether we all learn those lessons, and then, whether we take the appropriate actions in time.

The bipartisan Levin–Coburn Report issued by the US Senate found 'that the crisis was not a natural disaster, but the result of high risk, complex financial products; undisclosed conflicts of interest; and the failure of regulators, the credit rating agencies, and the market itself to rein in the excesses of Wall Street (Levin and Coburn, 2011).

The majority of individuals running the developed world's private and public financial institutions and the keepers of conventional wisdom in New York, London and other major developed country financial centres were virtually unanimous in acting as though they had repealed the laws of economics, and that growth in investments and profits could continue indefinitely, no longer restricted by macroeconomic conditions or local realities.

In a strikingly similar way, those individuals running the world's developed economies now seem to think that economic growth is also not restricted by physical or social conditions. The increasing inequality even within developed countries– so ably spotlighted by the 'occupy' movement is just one inevitable result of the assumption that nations can carry on consuming at ever-escalating rates, even as another three billion people join the population by 2050. For if those people start to consume in a similar way to populations in Europe or North America – as they would have every right to desire to – we'd all be striving to win increasing pieces of the same single, yet somehow ever-expanding, economic pie.

The assumption that such a world view was viable was challenged as long ago as 1972, by the Club of Rome's ground-breaking *Limits to Growth* (Meadows *et al.*,1972).The report analysed the available 'natural capital'– the total geophysical and ecological resources of the planet – and the rate of human consumption. It warned of the need to change global consumption patterns and factor in the depreciation of the Earth's resources, or to risk their devastating depletion.

Today those concepts seem obvious. But at the time – only 40 years ago – the report was attacked by governments, industry, labour organizations and NGOs alike, although of course for very different reasons.

Business and traditional economic leaders had taken it as an article of faith that the model of capitalism they followed would allow for – and encourage – an ever-expanding cycle of production and consumption. Governments, well aware that such growth provided precisely the material wealth that vast majorities of their constituents wanted, had no interest in rocking the theoretical boat. Trade unions, fighting for larger pieces of the economic pie, were pleased to support a model that saw that pie growing larger and larger – even though they ended up getting fewer and fewer additions to the pie over time.

Even many environmentalists, NGOs and more radical economists were concerned, because they thought that the report did not go far enough. They worried that its prescriptions, if followed, would help solidify a form of capitalism that they felt always left behind the poorest, and that dangerously impacted on the well-being of global ecosystems and of other species. Theirs was very much a minority viewpoint.

Commenting on *Limits to Growth*, the economist Matthew Simmons wrote: 'The most amazing aspect of the book is how accurate many of the basic trends extrapolations ... still are some 30 years later' (Simmons, 2009). An example is the book's suggestion that the gap between rich and poor would grow, in contrast to mainstream economic theory that it would decrease and the middle class expand, or at worst remain stable. In reality, that gap has grown ever wider – particularly in the past two deregulation-fuelled decades. To many, 40 years ago, it seemed impossible that human beings might increase their population and at the same time their economies to the point where they would alter Earth's natural systems. But experience with the global climate system and the stratospheric ozone layer has proven them wrong. It would be important to note though that growth has allowed very significant numbers living in poverty to diminish – mainly because of China.

Something like 40 per cent to 15 per cent of the world's population in the same time period are now judged to be living in poverty. The point is that our growth path is not sustainable.

This does not mean that growth must cease. Rather what is required is a fundamental change in the nature and content of growth. Just as in the growth of humans, the principal need is for healthy physical growth from birth to maturity. Thereafter, growth continues in the non-material dimensions of life – its intellectual, cultural and spiritual development. Indeed if its material growth continues after maturity it becomes unhealthy and ultimately unsustainable. The same is true of economic growth. The value of so many of the main products that contribute to GDP today – compact discs, DVDs, videos, etc. – depend to only a limited extend on their material content. Making the transition to a new mode of growth will mean there be no 'limits to growth'.

The current global economic system and its governance have not been effective in addressing the emerging problems that countries are now facing. Instead, the drivers of the global economy amplify and exacerbate them. As some fossil fuels become scarcer, greater efforts are being made to develop more marginal sources – with potentially more severe environmental and pollution impacts. The challenges ahead in the area of energy provision are huge.

The financial crisis

The parallels between the world's ecological crises and the financial crisis are clear. Both are products of a failed economic and social system. In the 1970s, 1980s and 1990s the banks and financial institutions of developed nations privatized the global system's gains and socialized its losses. The result was incredible profit for themselves when things were good, and incalculable losses for everyone else when things went bust.

We have been doing the same with the planet's natural capital. Humanity is increasingly living beyond its means – we currently consume 50 per cent more natural resources than the Earth's ecosystems can replenish. And that is before adding another one billion people by 2020 and three billion by 2050 (WWF, n.d.).

Using the ecological footprint model, Andrew Simms and David Woodward estimated that to lift the income of the world's poorest to not a mere US$1 a day, but to a still meagre US$3 a day, would require the equivalent natural resources of 15 planets like Earth (Woodward and Simms, 2006). As populations rise and expectations grow or at least remain the same, we will face a series of ecological crises in the years to come. The present lifestyles of OECD nations and newly industrialized countries (NICs) are drawing down the ecological capital from other parts of the world, and from future generations. We are increasingly becoming the most irresponsible generation our planet has seen.

The past 30 years have been characterized by rampant capitalism, pursuing limitless economic growth with little or no regard for the natural resource base upon which such wealth is built – at the expense of both society and the environment. An example of that cost can be seen in Iceland, where the either criminal or simply naive actions of a few banks – which copied with gusto the highly-leveraged, speculative model of their counterparts in the US – left a debt amounting to US$330,000 for every man, woman and child in the country (Lewis, 2009). Underlining the collapse is that the previous year (2008) Iceland had been number one on the United Nations' Human Development Index. Describing the degree to which the country had bought into the 'casino capitalism' atmosphere emanating from Wall Street, an IMF official visiting Iceland in 2008 (after the crash) is quoted as having said: 'You have to understand … Iceland is no longer a country. It is a hedge fund' (Lewis, 2009).

The people who managed our banks and financial institutions in the developed countries have failed us. So have our politicians. They have not held accountable those who were responsible for the execrable mess. How many of the most culpable bankers and investors ended up in jail? Not many. Nor are they yet putting into place systems that would ensure that it all does not happen again. As Joseph Stiglitz put it about those who designed the Obama Administration's rescue package, they are 'either in the pocket of the banks or they're incompetent' (Stiglitz, 2009).

An obvious lesson that was frequently agreed in principle was that no institution should become too big to fail. Even from candidates in the US Republican presidential primary debates, one could hear this call: 'With respect to the banks that are too big to fail … six institutions [today] are equal to 65 per cent of our GDP, $9.4 trillion … I say we need to right-size them' (Huntsman, 2011).

Unfortunately, little has been done in practice to assure such limits on the size of financial companies. And the candidate speaking was an anomaly in a political party that prefers to repeatedly blame the failures of de-regulation on a claim that banks and businesses were regulated *too much*.

One of the welcome major political developments of 2011, therefore, was the 'occupy' movement, which literally started on Wall Street and has spread across the developed world. It has targeted the massive disparity of wealth that now exists within countries, trying to catalyse a recognition that the last 20 years have not seen a trickle down of wealth, as had at least been assumed by critics, but instead a trickle (or perhaps more accurately a rush) *up*. The torrent of wealth flowing to the already-rich has been often at the expense of protection for ecosystems, the economic security of millions who had achieved the middle class, and the social support that provided some rays of hope for the poor.

Today, the principal goal of our economy must be to improve the lives of all of the world's people, particularly the one to two billion who live in a cycle of dire poverty, illiteracy, environmental scarcity and ill health. The challenge, of course, will be to accomplish that without compromising the resources of the planet itself. An economy that integrates sustainable development principles with responsible capitalism should be able to produce enough wealth to meet the needs of people in all nations, equitably and efficiently. We need to develop a roadmap to set nations on the path to a new 'economy' that is sustainable, equitable and accessible to all. To enable this to happen will require strong governance mechanisms and institutions.

The actual reaction to the crisis

Instead, actions by governments have ranged from the moderately ambitious to the tepid. Many governments acted to save the banks within their countries that had caused this problem by the irresponsible way they had behaved. The recovery packages were mixed, and some countries recognized the opportunity to move towards a greener economy. HSBC produced an analysis of how green the recovery packages were. Five key questions need to be asked about the relationship between the current crop of economic recovery plans and climate change:

1   Are plans allocating enough resource to the green stimulus?
2   When is the green stimulus likely to materialize?
3   How green is the green new deal?
4   How many jobs will be created in the short and medium term?
5   How effective is the green stimulus at mobilizing private investment?

It is perhaps easy to forget that in 1992 an assumption by so many attending Rio was that the price tag for the transformation of our economies to a 'greener one' was going to come from the peace dividend of the breakup of the former Soviet bloc.

Estimates differ, but recently it has been said that money invested in clean energy is estimated to create twice as many jobs per dollar invested compared with traditional fuel-based energy (Pernick *et al.*, 2010). This is indicative that the stimulus actually builds the base for sustained employment in low-carbon industries in the upturn. HSBC's review of the recovery packages found that governments allocated over US$500 billion to green packages:

> Since the beginning of the year [2009], the 'green stimulus' has grown from US$430 to over US$470 billion, with China and the USA in the vanguard, driving investment in critical infrastructure such as rail, grids, water, buildings and renewables.
>
> (HSBC, 2009)

As we write the book we don't yet have the information on how much actually was spent. We do know that the proposed US green bank did not happen. Looking at the report, it has some clear information on the percentage of the recovery packages that was invested in green activities:

- South Korea: 80.5 per cent
- China: 34.3 per cent
- Norway: 29.7 per cent
- France: 21.2 per cent
- Germany: 13.8 per cent
- USA: 12 per cent
- UK: 10.6 per cent

- Spain: 5.8 per cent
- Italy: 1.3 per cent.

Clear leadership is being shown in Asia, and what is disappointing is some of the responses by European countries who talk the talk on the green economy, but when it comes to the 'rubber hitting the road', as we say in the UK, there weren't the numbers that there should have been. The US did increase to around 20 per cent after the second stimulus package by President Obama. Unfortunately projects were blocked by Congress or by individual states such as the rail system expansion. Even with the ones that happened there is no evaluation on 'how green they were in the end'.

The UK reduced value added tax (VAT) as a mechanism to give people more money in their pockets and then went on to gut public services to pay back for the mistakes of bankers and politicians. If they had invested that amount in green jobs that could have not only had a positive impact on the employment rate, but would have contributed to the value of the country; although as most Europeans have opted out of producing solar panels, it would have also helped China's economy.

As this crisis happened in the run up to the Copenhagen UNFCCC meeting, there was a tendency to focus on what contribution recovery packages might make towards low carbon economies; in particular a focus on how this could be a 'jobs rich' approach at a time of soaring unemployment in many developed countries. Recovery packages looked at schemes like retrofitting buildings and ensuring new builds were low carbon, as well as improving mass transportation. Bill Gates, then head of Microsoft, said that creating a green economy is 'the biggest economic opportunity in the country' (Llewellyn et al., 2008: 29). He wasn't the only one to recognize this. Companies, not just governments, were investing in green technology. Some see green technology as driving the fourth industrial revolution, following the information age (Henderson 2010).

In the period of 2010–2011, investment in clean technology increased fourfold to US$234 billion (Bloomberg New Energy Finance, 2012). There is a growing acceptance that the next wave of productivity and innovation will come from green and smart technologies – some not new but re-found– that enable us to live more effectively in our resource-depleted world. The move to green jobs will be a gradual process, but can be accelerated; meanwhile brown jobs will continue and perhaps become less brown. The quicker the move towards greener jobs, the better – but this move needs to be underpinned by some principles.

The green economy, Rio+20 and MDG+15

Governments' agreement on the 'green economy' theme occurred surprisingly early in the process of the General Assembly deciding to hold the Rio+20 Conference. The formal, typically clumsy phrasing –'The Green Economy, in the context of Sustainable Development and Eradication of Poverty' – was an attempt to balance competing views emphasizing the economic value of 'green' policies (primarily advocated by the EU, but also by Brazil) and the concerns that the principles of sustainable development not be obscured, that green policies not become a pretext for constructing trade barriers against poorer countries, and that 'green labelling' not be utilized improperly by businesses for 'green-washing' (positions advocated by developing countries, supported by many NGOs).

As we write this, the 2012 Summit is still to happen, but it presents the possibility of revitalizing the image of sustainable development and re-invigorating the movement towards sustainable economies. The focus on a 'green economy' provides significant opportunities to expand participation to major actors in the economic sector – multinational corporations, small and medium companies, IFIs, and trade-oriented intergovernmental organisations (IGOs) – but also presents significant challenges to properly define for governments and the public precisely what a green economy means.

While use of the phrase 'sustainable development' (and 'sustainable consumption and production') has been hindered by accusations from some rigidly pro-business advocates and the political right that it will intentionally limit growth – and also, frankly, by its awkward linguistic quality – the phrase 'green economy' evokes an open, environment-friendly, people-friendly and business-friendly reaction. The fundamental belief of the 'green economy' model is that it is possible to increase economic prosperity, increase the number of decent jobs, reduce poverty and build healthy, active communities by beginning a proactive, environmentally informed economic transition. The risk of it – as its critics passionately warn – is that the green economy opens the door to further corporate aggrandisement under a guise of 'green practice' and is used to establish conditionalities to further restrict developing countries' exports and trade. This has been the constant challenge from 1972 onwards. It will be the responsibility of governments at Rio – and the thousands of civil society organizations tracking them – to make sure that this does not occur. They must assure that sufficient principles of implementation are established, and effective mechanisms of compliance are available and enforced.

If that is achieved, the green economy model holds the potential to bring together the simultaneous needs for economic security and environmental protection, and to integrate the complex array of reforms in policy initiatives, regulatory actions, voluntary business standards and personal lifestyle behaviours (which has not happened in any significant way since 1992) necessary to achieve immediate-, mid-, and long-range transition to a sustainable global system. The reality is of course that just 'saying that we have to internalize externalities' is no longer enough, models need to be shown to work and politicians need to be persuaded to accept those changes. Of course 'business as usual' is also not an option.

But the green economy's greatest strength may be its influence as a communications tool – its ability to communicate that environmentally-friendly policy can be jobs-friendly policies, and to help people visualize a positive, attractive, achievable vision of what a green society and a green economy could look like. It can help sell the idea that green communities would be economically thriving, liveable, healthy places for individuals and their families to live. And to emphasize that life for all could be more pleasant in a safe, just and equitable world.

The simultaneous and essential goal of such a strategy must be to reach, explain and to convince broad numbers of the global public to understand the sustainability principles behind the green economy model; and to form powerful consumer and political constituencies that will reward legitimate positive actions by both businesses and governments – and will punish fraudulent attempts to claim 'green' practice.

With the MDG+15 Summit scheduled in 2015, the conference in Rio offers an opportunity to agree a three year programme of activities that can be launched in 2012, and can then be reviewed three years later within the term of office of most present prime ministers and presidents.

Green economic job creation and economy-boosting opportunities

One of the most carefully elaborated efforts to define the need for a green economy is the UNEP Green Economy Initiative. The UNEP report provides an extensive analysis and policy framework to chart the feasibility and parameters of an international transition to a green economy model (UNEP, 2011a).

The UNEP report shows that the green economy model is expected to generate an equal amount – or more – of growth and employment compared with the current business-as-usual scenario, and it outperforms economic projections in the medium and long range, while yielding significantly more environmental and social benefits. It demonstrates that 'the greening of economies is not generally a drag on growth but rather a new engine of growth; that it is a net generator of decent jobs, and that it is also a vital strategy for the elimination of persistent poverty'. The report also seeks to motivate the enabling conditions for increased investments in a transition to a green economy. Among the report's most impressive conclusions is that:

- Investing 2 per cent of global GDP annually (currently approximately US$1.3 trillion) between now and 2050 into ten key sectors (agriculture, fisheries, forests, energy, water, waste management, manufacturing, construction, transportation and tourism) can fast-start a transition towards a low-carbon, resource-efficient economy.
- A green economy avoids considerable downside risks such as greater water scarcity, the effects of climate change, and impacts on agriculture, fisheries and forestry.
- Greening the economy generates growth in general, and particularly in natural resource sectors (agriculture, fisheries, forestry, freshwater).
- It also produces a higher growth in GDP and GDP per capita. The report's green investment scenario achieved higher annual growth rates than a business-as-usual scenario within 5–10 years, characterized by a significant decoupling from environmental impacts with the global ecological footprint to bio-capacity ratio projected to decline from a current level of 1.5 to less than 1.2 by 2050 – as opposed to rising beyond a level of 2 under business as usual.

- The green investment scenario is projected to reduce energy-related CO_2 emissions by about one-third by 2050 compared with current levels, with atmospheric CO_2 held below 450 ppm by 2050.
- A green economy can contribute to poverty alleviation, both because of job creation and due to benefit flows from natural resources that are directly utilized by the poor, particularly in low-income countries, where ecosystem goods and services are a large component of the livelihoods of poor rural communities and provide a safety net against natural disasters and economic crashes.
- New jobs will be created in a green economy transition, which, over time, will exceed the losses in 'brown economy' jobs. This is particularly notable in the agriculture, buildings, energy, forestry and transport sectors.
- Reforming costly and harmful subsidies in all sectors will open fiscal space and free resources for a green economy transition. Removing subsidies in energy, water, fisheries and agriculture sectors, alone, would save 1–2 per cent of global GDP a year.
- Using financial incentives, taxes and tradeable permits to promote green investment and innovation is essential, but so is investing in capacity-building, training and education, and strengthening international governance and global mechanisms that support a transition.
- Financing required for a green economic transition, while substantial, is less than one-tenth of annual global investment (2 per cent of global GDP is required under the UNEP model, while total gross capital formation was 22 per cent of global GDP in 2009).
- Globally, new investment in clean energy was expected to reach a record high of US$180–200 billion for 2010, up from US$234 billion in 2010–2011. However, clean energy growth is increasingly driven by developing nations (e.g. China and Brazil) – so US business risks squandering a leadership role in a potentially booming economic sector (non-OECD countries' share of global investment in renewables rose from 29 per cent in 2007 to 40 per cent in 2008).

Principles for a green economy

The UNEP definition is only one of several. Still at an early stage of policy definition, its very amorphousness serves in some ways as an advantage. There is not a generally accepted definition for the green economy. The United Nations Environment Programme has produced one version for its Green Economy Report (Box 12.1).

At this point, a long debate about precise definitions may not be productive. More useful might be a formulation of fundamental principles that might be used for guiding the economy on a sustainable path.

An approach advocated by the Earth Charter Institute, Stakeholder Forum for a Sustainable Future, and Bio-Regional, utilizes the Stockholm, Rio and Johannesburg Declarations, the Earth Charter, One Planet Living, the trade union-sponsored Just Transition to suggest a set of composite principles (Stakeholder Forum, 2011a):

Box 12.1 **Principles for a green economy**

1. *Equitable distribution of wealth:* Promote the equitable distribution of wealth within nations and among nations, to reduce disparities between rich and poor, and achieve social and economic justice, within a sustainable and fair share of the world's resources and leaving sufficient space for wildlife and wilderness.

2. *Economic equity and fairness:* Guided by the principle of common but differentiated responsibilities, create economic partnerships that would transfer substantial financial and technological assistance to less developed countries, to help minimize the gap between the developed and developing world and support the environmental sustainability of both.

3. *Intergenerational equity environmental resources and ecosystems* must be carefully managed and safeguarded so as to enhance the value of environmental assets for future generations, thereby equitably meeting their needs and allowing them to flourish.

4. *Precautionary approach:* Science should be utilized to enhance social and environmental outcomes, through the identification of environmental risk. Scientific uncertainty of environmental impacts shall not lead to avoidance of measures to prevent environmental degradation. The 'burden of proof' should lie with those claiming that there will not be significant environmental impacts.

5. *The right to development:* Human development in harmony with the environment is fundamental to the achievement of sustainable development, so that individuals and societies are empowered to achieve positive social and environmental outcomes.

6. *Internalization of externalities:* Building true social and environmental value should be the central goal of policy. To this end, market prices must reflect real social and environmental costs and benefits, so that the polluter bears the cost of pollution. Tax regimes and regulatory frameworks should be used to 'tilt the playing field', making 'good' things cheap and 'bad' things very expensive.

7. *International cooperation:* The application of environmental standards within nation-states must be undertaken in a cooperative manner with the international community, based on an understanding of the possible impact on the development potential of other states.

8. *Environmental measures relating to trade* should avoid unfair protectionism, but overall should ensure that trade supports sustainable resource use, environmental protection and progressive labour standards, promoting a 'race to the top' rather than the bottom.

9. *International liability:* Acknowledging that actions within national boundaries can cause environmental impacts beyond national jurisdictions,

continued...

Box 12.1 continued

requiring cooperation in the development of international law that allows for independent judicial remedies in such cases.

10. *Information, participation and accountability:* All citizens should have access to information concerning the environment, as well as the opportunity to participate in decision-making processes. To ensure that environmental issues are handled with the participation of all concerned citizens, institutions at all levels (national and international) must be democratic and accountable, and make use of tools that enable civil society to hold them to account. In this regard, the access to justice by citizens for redress and remedy in environmental matters is a cornerstone of enhancing accountability.

11. *Sustainable consumption and production:* Introduce sustainable production and consumption with sustainable and equitable resource use. Reduce and eliminate unsustainable patterns of production and consumption (i.e. reduce, reuse, and recycle the materials used, acknowledge the scarcity of the Earth's resources and implement activities accordingly).

12. *Strategic, coordinated and integrated planning* to deliver sustainable development, the green economy and poverty alleviation. An integrated approach must be adopted at all levels to expedite the achievement of socio-economic and environmental sustainability through strategic planning with civil society and stakeholders, and across all relevant government departments.

13. *Just transition:* There will be costs in making the transition to a low-carbon, green economy in the pursuit of sustainable development. Some states and actors are better able to bear those costs than others and are more resilient to transitional changes. In the process of change, the most vulnerable must be supported and protected – developing countries must have access to appropriate financial and technical assistance, citizens and communities must also have access to new skills and jobs.

14. *Redefine well-being:* GDP is an inadequate tool for measuring social well-being and environmental integrity. Many socially and environmentally damaging activities enhance GDP – such as fossil fuel exploitation and financial speculation. Human well-being and quality of life, and environmental health should be the guiding objectives of economic development.

15. *Gender equality:* Gender equality and equity are prerequisites to the transition to a green economy and the achievement of sustainable development. Women have a vital role to play as agents of change for environmental management and development – their actions must be rewarded accordingly and their skills enhanced.

16. *Safeguard biodiversity and prevent pollution* of any part of the environment: Protect and restore biodiversity and natural habitats as integral to development and human well-being, and develop a system of governance that protects the resilience of ecosystems to prevent irreversible damage.

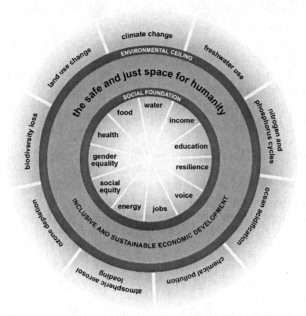

Figure 12.1 A safe and just space for humanity to thrive in: priorities for Rio+20
Source: Raworth (2011). The social foundation is based on governments' priorities for Rio+20. The environmental ceiling is based on Rockström et al. (2009)

Guiding all of these principles is the premise that the transition to a green economy will only be politically acceptable if it is pursued in an equitable way, both within and between countries. The green transition must be a just and fair one.

Social issues for a green economy

Addressing the social issues, Oxfam researcher Kate Raworth has proposed to counterbalance the evolving narrative around nine planetary boundaries, proposed by Johan Rockström of the Stockholm Environment Institute, with the development of social boundaries. She proposes a doughnut as the visual, with social boundaries at its core, and a safe and just operating space between them and the planetary boundaries on the outside part of the doughnut (Raworth, 2011).

Early discussions are being conducted on this as we write the book, but it looks like a possible area to develop in time for the MDG review in 2015. She suggests that they might be derived from human rights (e.g. a right to water) and human capabilities, and the MDGs themselves. This would be seen in the development of what is being called sustainable development goals (SDG). The SDGs were proposed by many people but formally tabled for governments to respond to by the governments of Colombia and Guatemala at an international workshop on the Institutional Framework for Sustainable Development Governance (IFSD) in Solo in Indonesia in June 2011. They may take the supplement, integrate or even supplant the MDGs after the 2015 MDG implementation deadline has passed.

The approach to creating new green jobs has been supported by the global trade union movement through the International Trade Union Confederation. Often over the last 20 years the discussion on the environment has often seemed to be pitching the environment movement against the trade unions movement, but no more. The anti-WTO protests in Seattle in 1999 began a process of integrating the goals of environmentalists, development organizations, and trade unions:

> The international labour movement agrees with the urgency of moving towards a resource efficient, low-carbon economy, prioritizing renewable resources over non-renewable, implementing a life-cycle approach to products and better integrating environmental benefits and costs in production and consumption in order to protect and, where possible, restore ecosystems.
>
> (International Trade Union Confederation, 2011)

The important issue here is whether the global or national or local transition is going to be done in a way that is people-friendly and that creates decent green jobs: a just and fair transition.

ILO's involvement with UNEP and the ITUC on a report (UNEP, 2008) produced in the framework of creating a green jobs initiative assembled evidence – quantitative and conceptual – on existing green jobs. That just transition should underline support for the ILO Convention known as the Social Security (Minimum Standards) Convention (102); it is the flagship of all ILO social security conventions, as it is the only international instrument, based on basic social security principles, that has established since 1952 minimum standards agreed worldwide for all nine categories of social security. These branches are:

- medical care;
- sickness benefit;
- unemployment benefit;
- old-age benefit;
- employment injury benefit;
- family benefit;
- maternity benefit;
- invalidity benefit; and
- survivors' benefit.

Any future green economy should see states that have still to ratify the ILO Convention 102 – such as the US – to do so as a commitment to a just transition. There has been great progress now with more than 40 developing countries with social safety nets. What is missing in the dialogue is how to help small- and medium-sized enterprises (SMEs), which is where most jobs are created in both developed and developing countries.

Key policy areas that should be considered

We have identified a number of areas which would help us move towards a more just and sustainable economy. These would include:

- government actions – taxation and subsidies;
- cities and the green economy;
- sustainable consumption and production;
- Convention on Corporate Social Responsibility and Accountability and Voluntary models;
- financial markets and the green economy;
- new economic indices;
- national wealth accounting.

Taxation, funds and subsidies

This section looks at three main areas:

- taxation and subsidies;
- financial transaction tax;
- climate funds.

Taxation and subsidies

The national fiscal policies dealing with taxation and subsidies should be aligned to support sustainable development. To support the move to greener jobs, governments can adopt a number of fiscal instruments. In the area of taxation a strong move away from taxing income to taxing resource depletion and pollution would be an important step. But while doing that the funds would still cover the same social infrastructure. This would act as driver to resource reduction and life cycle approach. It would reward companies that behaved in a responsible way.

In a market economy that drives the processes of globalization, the market provides the signals that motivate sustainable development. This means shifting taxes from products and practices that are environmentally and socially beneficial to those that are least harmful; in effect, getting the prices right. No nation can do this alone without disadvantaging its own economy; it has to be done through international agreements. There is a lot of room for individuality in the manner in which we administer these nationally, but it can only be effectively done within an internationally agreed framework.

In the area of subsidies, the removal of 'harmful' subsidies on fossil fuels and agriculture in particular is estimated to be in the order of US$750 billion a year (UNEP, 2011b). This money could be invested to support the development of new renewable energy development as well as supporting other sustainable development practices that generate jobs. The refurbishment of most of the housing stock and offices in the developed countries could be supported through local green bonds, which might be underwritten by new money from reducing harmful subsidies.

Financial transaction tax

A financial transaction tax (FTT) has been on the table since John Maynard Keynes suggested it in 1936 as a small transaction tax, to avoid excessive speculation. He was worried about the way that speculators could have a disruptive impact and would become dominant if not restrained: 'Speculators may do no harm as bubbles on a steady stream of enterprise. But the situation is serious when enterprise becomes the bubble on a whirlpool of speculation' (Keynes, 1936).

The idea had been repeatedly advocated by many NGOs as a vehicle to generate funds for green initiatives and by governments that at different times included France (2009), Germany (2009), Spain (2004), Austria (2011) Belgium (2004), Finland (2000), Brazil (2009), Argentina (2007), South Africa (2011), Cuba (2001) and Venezuela (2001). Yet it failed to gain traction in intergovernmental negotiations. It has now taken on new life in the fierce debates over how to stabilize the Eurozone economies. The European Commission has suggested a FTT could be introduced within the Eurozone.

This to date has been opposed by the major economies of China (due to the burden on domestic banks), India (burden on domestic banks), US (has supported it but could not get it through Congress), UK (unless global) and Canada (against taxes), setting up what appears to be a whole new 'big economy versus everybody else' polarity that includes both traditional and newly emerging economic super powers on the side of opposing any slowdown in hyperactive global trade-based economy.

James Tobin in a 1972 Janeway Lecture at Princeton suggested that the tax be a penalty on short-term financial round-trip excursions into another currency. This was shortly after the Bretton Woods system effectively ended their stabilizing of currencies.

The Chicago Political Economy Group (CPEG) estimates that such a tax would bring in the region of between US$750 billion and US$1.2 trillion per year in the US alone:

> FTT would be a small tax on all trading in stocks, currencies, and debt products such as treasury bills and bonds (and futures and options contracts on all of these). The amount of the tax would be quite small. For example, a tax of $1 on every $400 of stocks traded (0.25%; one-quarter of one percent) and $1 on every $800 of currency and debt trading including derivatives (0.125%, one-eighth of one percent).
>
> (CPEG, 2010)

The question is what the funds would be used for. With the present economic situation the tax may be used to replenish government funds as opposed to development aid, and perhaps now for helping fund a transition to a more sustainable economy. The US in particular might be persuaded about this to address the US deficit; this might be true for other countries too. The experience of Sweden introducing it unilaterally wasn't positive and there are clear lessons to learn there, including the level the tax was introduced at.

Climate funds: incentives and means

The Clean Development Mechanism (CDM) process makes substantial contributions to the transition by industry to sustainable, scientific development in providing the incentive and means to employ the latest technologies and develop their own technological capacities. It also makes important contributions to health and safety. China, which continues to rely on coal for some 70 per cent of its energy, can use CDM funding to capture and utilize methane, which is the source of so many tragic mine accidents that have happened in China and is one of the main contributors to greenhouse gas emissions.

In China, purchases of Certified Emission Reductions (CERs) have concentrated on the larger projects which produce substantial credits. But as 'low-hanging fruit' becomes less available; purchasers are beginning to move towards smaller projects, some of these involve packages of projects, such as the production of low-cost and low-emission stoves and other projects that can meet grassroots needs while enabling them to benefit from CDM funding.

With the immense increase in public awareness and political attention to the climate change issue, arising largely from the latest scientific findings of the Intergovernmental Panel on Climate Change and the initiatives of former US Vice President Al Gore, which were recognized by their award of the 2007 Nobel Peace Prize, there has been a rapid increase in the response of the financial community to the opportunities created by the carbon market. New carbon funds are proliferating and there are few investment banks or financial institutions that are not moving into the field.

The World Bank has taken a lead in promoting the utilization of the CDM process to finance development projects which produce significant emission reductions in developing countries. More controversially, it has become a major competitor, with private funds for the purchase of CERs.

While CDMs and emissions trading must be seen as only a partial and a still imperfect response to the challenge of climate change, they nevertheless are the main game in town in terms of new and expanding opportunities for the financial community, that at the same time will contribute significantly to China's commitment to scientific development and a sustainable, harmonious society.

The new green fund in the climate convention agreed to at the Durban UNFCCC CoP (2011) has now been agreed to; when and how it will be set up is still open for discussion and agreement.

Cities and the green economy

Both existing and new cities must be green and can generate green jobs. Eco-cities must be planned and developed as complex systems using the most sophisticated of networking, computer, materials, building and energy technologies. The economic and social conditions of each city will be distinctive and must be understood in terms of the intensity, density, use of space and transportation systems which link them both within the city and beyond. This requires a rational distribution of their population and the relationship between where people live and where they work.

Civilization as we know it began in the cities and cities continue to be the centres of our civilization. Most environmental problems originate in the cities and in the impacts they have on their hinterlands, and most of the opportunities for improving the environment and the capacities to do so are centred in the cities. The critical role of the cities is becoming even more important as the processes of urbanization concentrates more and more people in cities. While this creates growing problems and challenges it also provides opportunities for effective management of environmental problems.

Compactness is important. The contemporary city should not expand in green fields, opting for reuse of existing degraded or abandoned built areas. A compact city reduces the need for private transport, lowers public costs for services and reinforces the sense of community and social cohesion. Compact cities can also offer more opportunities to women and young people. Densely populated areas are more sustainable than sprawling urban agglomerates and provide more efficient services. According to an OECD study, congestion costs amount to 1–1.3 per cent of GDP. On the other hand, density makes productivity jump by 6 per cent.

Today in developed countries, cities compete with each other to attract residents. Residents are tax payers and contribute to the city balance, but also create jobs. According to the estimates, every 1,000 residents create 25 new jobs in a local community. The location of industry is of special importance – in its relationship to transport, accessibility to employees and, of course, the primary contribution it must make to environmental sustainability and reduction of carbon emissions. To meet these monumental challenges, cities must be understood and developed not merely as places but as urban systems.

The C40 Cities Climate Leadership Group has estimated that 75 per cent of us will live in urban areas by 2050. China is committed to building 500 new cities each with a minimum population of a million or more people. These are all designed to be 'green cities' which will relieve the pressures on China's existing, rapidly expanding cities and contribute to the development of a green economy.

By 2030 urban areas in the developing world will triple. In the next 25 years in China, 300 million people will move in cities that have yet to be built. The United Nations Population Fund (UNFPA) has calculated that by 2050 Asia's urban population will double, from 1.7 to 3.4 billion, while Africa's will triple from 400 million to 1.2 billion. The footprint of our cities will be critical to our sustainability. In March 2011 UN-Habitat brought out a report, *Cities and Climate Change* (UN-Habitat, 2011), which estimates that, while covering only 2 per cent of global land area, cities account for 70 per cent of greenhouse gas emissions. The report suggests that 8 per cent of global CO_2 comes from buildings, 23 per cent from transportation and 19 per cent from industry. An approach to address all three of these will create local jobs.

Cities also consume 75–80 per cent of world resources, and produce 70 per cent of global waste. The ten biggest megacities reach a total GDP greater than a whole nation like Japan, and the 50 major cities have a GDP larger than China and second only to the USA. Cities will be critical in our resource-limited world of the future, but work now can create local jobs through refurbishment of buildings, through creating small and medium urban agriculture production and by creating local farmer markets to sell the products.

The European Parliament in May 2010 voted a resolution asking for all new buildings to be carbon neutral by 2020 (and even sooner for public buildings – by the end of 2018). In the area of refurbishment a number of initiatives could move this forward. The UK government, through their CRC Energy Efficiency Scheme, has a mandatory process aimed at improving energy efficiency and cutting emissions in large public and private sector organizations. These organizations account for 10 per cent of the UK emissions. This could be copied in a number of developed countries (DECC, 2011).

The European Union has successfully launched the Covenant of Mayors, a programme to enrol cities in a voluntary process to reduce CO_2 emissions by at least 20 per cent by 2020 through the enforcement of city energy plans. Since 2009 more than 3,300 European local authorities have signed the Covenant and 1,100 local sustainable action plans have been submitted. The Mexico City Pact that was launched by ICLEI and C40 in November 2010 extends the Covenant's principle of voluntary commitment to cities outside Europe.

In Cuba we have a very good example of what can happen. The Soviet Union withdrew its financial support for cheap oil for Cuba, which then had to find ways of reducing their energy use while feeding their population. The threat of serious food shortages was overcome in five years (Novo and Murphy, 2001). Urban gardens were to provide 60 per cent of the vegetables eaten in Cuba (Simms, 2009).

Trees and other plants in cities and towns can have a significant impact towards removing carbon from the atmosphere. It has been estimated if a city of around 300,000 is planted with trees then it can absorb around 12 per cent of greenhouse gases (Davies et al., 2011).

There are some great examples of how to transform cities to become more sustainable pioneered by ICLEI, United Cities and Local Government (UCLG) and other local government associations. These could be brought together and shared through a new UN Conference on Human Settlements.

Sustainable consumption and production

In an odd way, what we are really talking about is going back to the future. The ancients saw no division between themselves and the natural world. They understood how to live in harmony with the world around them. It is time to recover that sense of living harmoniously for our economies and our societies.

Not to go back to some imagined past, but to leap confidently into the future with cutting-edge technologies, the best science and entrepreneurship has to offer, to build a safer, cleaner, greener and more prosperous world for all. There is no time to waste.

(Ban, 2011)

Chapter 4 of Agenda 21 was an attempt to seriously address consumption patterns. It tried to promote patterns of consumption and production that reduce environmental stress, while at the same time meeting the basic needs of humanity. It also started a process (now 20 years old) to develop a better understanding of what is the role of consumption and how to move to more sustainable consumption patterns.

In the first five years the government of Norway played a significant role in hosting workshops to address this. In 1994 Norway hosted the Oslo Roundtable on Sustainable Production and Consumption (SCP), where they put forward a definition of SCP:

The use of services and related products which respond to basic needs and bring a better quality of life while minimizing the use of natural resources and toxic materials as well as the emissions of waste and pollutants over the life cycle of the service or product so as not to jeopardize the needs of future generations.

(Prinet, 2011)

The 1997 five-year review agreed to a new work programme, which saw poverty and sustainable consumption and production as cross-cutting themes that should be addressed each year on all topics. The need to change consumption patterns is clear. SCP is a key instrument for the achievement of progress towards sustainable development.

In the run up to the WSSD, Roper Green Gauge did a survey on consumption patterns in the US. They found that there were five segments of consumer. These were:

- *True blue greens* (major green purchasers and recyclers) 11 per cent
- *Greenback greens* (will buy or give green, won't make lifestyle changes) 5 per cent
- *Sprouts* (who care but would only spend a little money to buy green) 33 per cent
- *Grousers* (who see the environment as a problem but someone else's) 18 per cent
- *Basic Brown* (who essentially don't care/won't care) 31 per cent

With a huge challenge by the most unsustainable and consuming nation in the world, the JPoI describes SCP as one of the three overarching objectives and essential requirements for sustainable development, together with poverty eradication and the protection and management of the natural resource base. Chapter III of the JPOI calls for actions to:

... encourage and promote the development of a 10-year framework of programmes (10YFP) in support of regional and national initiatives to accelerate the shift towards SCP to promote social and economic development within the carrying capacity of ecosystems by addressing and, where appropriate, delinking economic growth and environmental degradation through improving efficiency and sustainability in the use of resources and production processes and reducing resource degradation, pollution and waste.

(United Nations, 2002b)

What became known as the Marrakech Process delivered its plan for discussion at the UN CSD in 2010, and for the second time in its history the UN CSD failed to agree anything.

There are a set of barriers to addressing sustainable consumption. These include the following factors:

- fragmentation of sustainability initiatives at all scales;
- inertia in the system;
- paralysis in front of tremendous complexity;
- need for political leadership;
- skewed markets resulting from misplaced subsidies, the externalization of costs, short-term profit seeking, a non-level playing field and entrenched interests;
- misunderstanding around the seriousness of issues;
- influence of powerful interest lobbies;
- inability to effectively apply a systems perspective to problem solving; and
- too much emphasis on efficiency as a means to reduce material and energy throughput when historically, lower prices brought about by gains in technology and efficiency have led to overall increases and bigger footprints when an overall cap on resource use is not established.

(Prinet, 2011)

It's against this backdrop of very low political interest that some industry and some government actions nonetheless focused on reducing production costs through initiatives, such as energy efficiency, sustainable procurement, greening supply chains, waste minimization and the 3Rs (recycle, reuse, reduce), and life-cycle impact analysis.

There have also been changes in consumption through personal action. In the US, for example, there has been the growth of local farmers markets: in 2006 there were 4,385 farmers markets; this increased to 5,274 by 2009, perhaps reflecting the movement for healthier foods and also a concern about food miles. As another example of personal action, there are now many personal calculators on the web to work out your waste production, your water consumption or your energy use.

Unfortunately, without the active participation of national governments, they will not allow individuals to make as much of an impact as they could. While opponents often argue that personal behaviour cannot be effectively modified by government regulation, the facts indicate precisely the opposite. One main example is smoking. In the last 20 years, in most developed countries, taxation, education and banning from public places have had a profound impact: from 2005 to 2010, smoking decreased in the US from 23.4 per cent to 19.3 per cent among adults over 18 years of age (MMWR, 2011).

There are a number of policy instruments that can be used to change consumption and production patterns:

- *Regulatory instruments:* Norms and standards, bans, caps and emission limits, building codes.
- *Economics instruments:* Fiscal or market-based instruments (subsidies, taxes, user fees, green procurement, bonus systems, emissions trading, etc.), investments in infrastructure that supports sustainable consumption and production.
- *Participatory instruments:* Urban planning, public dialogues, policy discussions, participatory budgets.
- *Voluntary instruments:* Unilateral commitments, negotiated agreements, voluntary setting of targets, voluntary certification schemes.
- *Information instruments:* Eco-labels, information centres, websites, demonstration projects.
- *Educational instruments:* Training, integration of whole systems thinking in formal and informal education, research.
- *Organizational/institutional arrangements:* Cooperation schemes, public–private partnerships, agreements between different levels of government.

(Prinet, 2011)

At the global level, agreeing on SDGs or reinvigorating national councils for sustainable development could play a significant implementing role. A discrete intergovernmental body for SCP, similar to UN Water and UN Energy, would integrate the whole UN system together in promoting SCP internally and externally in all development projects.

Even if population increase is unavoidable, and expansion of prosperity is achievable, the form that prosperity takes will be the decisive factor:

The big issue on the planet is not just the fact that we're going from 6.7 billion people today to 9.2 billion. It's the number of people living an American lifestyle, eating American-sized Big Macs, living in American-sized houses, driving American-sized cars, on American-sized highways.

(Friedman, 2006)

Convention on corporate social responsibility accountability

One of the missing components from Agenda 21 was an effective mechanism to address the role of transnational companies (TNCs) in a more globalizing world. A proposed 41st chapter of Agenda 21 had been put forward in 1991 by the UN Centre for Transnational Corporations (UNCTC). The draft chapter, titled 'Transnational Corporations and Sustainable Development', contained many good suggestions on sustainable development and TNCs in the areas of:

- global corporate environmental management;
- risk and hazard minimization;
- environmentally sounder consumer patterns;
- full cost environmental accounting; and
- environmental conventions, standards and guidelines.

The chapter called for actions such as:

56 (d) Address in future environmental instruments the rights and responsibilities of transnational corporations;

57 (b) Establish or strengthen a regulatory environment supportive of sustainable development.

(UNCTC, 1991)

Twenty years later, the possibility of the chapter has risen again. The government of Mexico, in its submission to the Rio+20 zero-draft text, proposed:

To endorse the proposal to include a new Chapter 41 of Agenda 21 on 'Transnational Corporations and Sustainable Development' which include measures such as:
   i Corporate Environmental Management, global patterns of production and sustainable consumption.
   ii Accounting systems that incorporate the total costs.
   iii The implementation of environmental conventions that promote the effective enforcement of environmental standards and guidelines and existing social responsibility such as ISO 26000 on social responsibility, ISO 14000 and ISO 9000 quality management systems in general and in particular environmental.

(UNCSD, 2011)

Governments, stakeholders and industry are all looking at how to address corporate sustainability in an intelligent and effective way. The last 20 years have developed some voluntary approaches to fill in the gap.

Voluntary models

Until now, cooperation on corporate responsibility has had to be almost entirely voluntarily. initiatives such as the Forest Stewardship Council and the Marine

Stewardship Council have achieved significant success, particularly in the area of certification. Such programmes have pioneered workable models. Again, however, without governmental regulation such efforts will simply be insufficient.

There have been broad initiatives, such as the UN Global Compact, set up in 1999 by UN Secretary-General Kofi Annan. While subject to intense criticism from a significant minority of NGOs who claim it enables 'green washing', the Global Compact has secured over 6,500 companies in support of ten principles addressing human rights, the environment, CEO water mandates, anti-corruption and labour standards. More recently there has been the UNEP Finance Initiative's (UNFI) work on areas such as the Principles for Responsive Investment (PRI) and the Principles for Sustainable Insurance (PSI).

An attempt by Friends of the Earth International and other NGOs in 2002 during preparations for the WSSD was to move forward with a convention on TNCs (Friends of the Earth, 2002). Ironically, it gained some industry support, but was opposed by trade unions, worried about the results of negotiating such a legal instrument with a highly conservative government, like that of the US under President George W. Bush. Friends of the Earth proposed a number of principles that could be embedded within such a convention. This included liability for directors for 'Corporate breaches of national social and environmental laws, and to directors and corporations of corporate breaches of international laws or agreements' (Griffiths, 2005). This resulted not in a convention but in the JPoI merely voicing support for some more voluntary action. NGOs then moved their efforts to the ISO process, a result of which was the ISO 26000 on Social Responsibility (2010), which takes into consideration seven principles:

- human rights;
- labour practices;
- consumer issues;
- community involvement and development;
- the environment;
- fair operating practices; and
- organizational governance.

Agreed by governments, stakeholders and industry associations, ISO 26000 provides an excellent starting point for input to a future convention. Although some companies had started to produce sustainability reports, at present more than 75 per cent of companies covered by Bloomberg do not currently disclose their sustainability performance.

A primary element of all proposals for a Corporate Responsibility Convention has been a requirement for publicly traded companies to report on their social and environmental impacts. This is now being championed by over 50 leading companies worth over US$2 trillion in investment, led by AVIVA, the largest UK insurance company, and sixth biggest in the world. Announced in the UN General Assembly in 2011, it calls for:

... Earth Summit delegates to develop a convention on Corporate Sustainability reporting. The Rio summit would ideally be a commitment to develop national regulations which mandate the integration of material sustainability issues in the Annual Report and Accounts.

(AVIVA, 2011)

They identify four principles:

1 *Transparency:* Companies should be required to integrate material issues within their report and accounts – or explain to the market why they cannot do this.
2 *Accountability:* There should be effective mechanisms for investors to hold companies to account on the quality of their disclosures, including for instance through an advisory vote of the AGM.
3 *Responsibility:* Board duties should explicitly include setting the companies values and standards and ensuring that its obligations to its shareholders and stakeholders are understood and met.
4 *Incentives:* Companies should state in remuneration reports whether the remuneration committee consider ESG factors which are of material relevance to the sustainability and long term interests of the company when setting remuneration of executive directors, aligning remuneration with the interests of shareholders and other key stakeholders, including customers and employees.

(AVIVA, 2011)

The coalition also calls for every company to present a Corporate Sustainability Strategy to a separate advisory vote at its AGM. This call originated in the finance sector but was supported in January 2012 by the University of Cambridge Programme for Sustainability Leadership, whose Patron is the Prince of Wales. The coalition endorsed the call for a convention:

Establish a holistic policy framework to sustain natural capital. It is essential that governments introduce policies that encourage companies, investors and regulators to assess natural capital impacts on company value chains, provide incentives for more sustainable products, and support effective collaboration across sectors between those reliant on the same ecosystems. As part of this framework, governments and companies must be required to disclose their environmental impacts. In support of this goal we urge that governments join with leading companies in support of a global convention on corporate reporting, based on principles of transparency, accountability and board responsibility, to develop national regulations that mandate the integration of material sustainability issues into public reporting to key stakeholders, including investors.

(University of Cambridge Programme for Sustainability Leadership, 2012)

This drew support from a number of large global companies indicating that support for CSR has spread to corporations far beyond those previously thought of as purely 'green', 'ethical' or 'sustainable'.

Finance, capital markets and the green economy

As opposed to the often destructive role of short-term speculators, there is a critical role for long-term investors that can be highly constructive if channelled effectively.

It is a factor that is frequently missing from the analyses of outright critics of the market system, but a creative and responsible utilization of the vast funds available for financial investment can be one of the few available vehicles that can realistically provide the means to enable the significant transition all agree is necessary.

The realization of a green economy in the twenty-first century will be set against a backdrop of an ever-changing global financial system that faces manifold challenges and emergent opportunities.

Markets remade: the backdrop for a green economy

The world is entering a period of fundamental change for finance and investment as economic power in the global economy is rebalanced, with much greater prominence for emerging economies such as the BRICs (Brazil, Russia, India and China).[1] The prospect of fundamental demographic changes in the years ahead also raises serious questions about how and where concentrated pools of capital captured in the developed economies savings structures will be invested in coming years. Equally, the extension of financial services to those 2.8 billion people 'with discretionary income who are not part of the formal financial system' (McKinsey, 2009), as well as the evolution of financial services for those at the base of the pyramid (Prahalad, 2005), is also a critical challenge.

The G20's (G20, 2009) stated goal of a stable, sustainable, and resilient global financial system, and the fundamental economic, social and environmental roles of the investment chain and processes of financial intermediation within that system, will also greatly influence the speed and ability to transition to a green economy. The financial crisis of 2007–2008 saw worldwide financial assets fall by US$16 trillion to US$ 178 trillion in 2008 (McKinsey, 2009) from their peak of US$194 trillion in 2007. The crash and subsequent severe global recession destroyed US$28.8 trillion (McKinsey Global Institute, 2009) in global wealth captured in equity and real estate values by mid 2009. During one week in October 2008, it is estimated that some 20 per cent of the 'value of global retirement assets were wiped out' (Authers, 2010). These value-destroying financial and economic crises[2] have come with greater intensity and higher frequency since the 'Black Monday' stock market crash of 1987 when the Dow Jones Industrial Average in the US lost more than 22 per cent in a day (Browning, 2007). Such events, with serious systemic implications, raise a fundamental question of whether the current financial system is 'fit for purpose' to deliver a low-carbon, resource efficient, inclusive green economy and, if not, what such a system should look like.

Importantly, new and complex questions about how future financial instability might be triggered and compounded by the myriad 'slow failures or creeping risks' (World Economic Forum, 2010), such as resource scarcity, climate change, threats to our biodiversity and ecosystems, the demographics of ageing populations, and the impact of chronic diseases have not been adequately explored by financial policy-makers or the finance and investment sectors. Such an exploration of financial stability and long-term systemic risks is only now just starting (IISD and UNEP FI, 2010).

The contradiction: finance and the markets

The financial services and investment sectors, representing the largest and most powerful set of globalized industries,[3] operate in a complex environment of evolving international policy norms that are not aligned with fragmented financial policy, legislation and regulation at the regional or national levels. There is no shortage of supervisory bodies or regulation around financial services and capital markets and, yet, in the run up to the crash of 2008 these systems failed. Within financial services and investment, an overwhelming focus on short term performance, often driven by intricate 'financial engineering', meant that in the run up to the crisis important parts of the industry created unimaginable levels of systemic risk that contributed to an almost complete meltdown of the global financial system. Since late 2008, the financial policy-making community worldwide has done three things: worked to prevent a severe recession becoming a global depression; sought to understand the fundamental reasons for the crash and subsequent global economic downturn; and have worked in 2010–2011 to develop and implement a policy environment and supporting regulatory framework that either heads off or minimizes the impact of future financial crises.

However, despite the fundamental questions being posed around the stability of the modern financial system, the period from 1990 to 2007, which saw an exponential growth in capital markets, also highlighted the power of public and private finance to allocate capital and evolve innovative forms of productive investment and financing that support economic growth, lifting hundreds of millions of people worldwide out of poverty. The financial assets controlled by our global capital markets and the worldwide financial services community increased by close to 350 per cent as a percentage of gross domestic product between 1980 and 2007, peaking at US$194 trillion (McKinsey, 2009).

Financial upside of the green economy

The market changes highlighted above have contributed to the emergence of a broad range of important developments that support the green economy agenda such as:

- the arrival of clean energy as a new asset class with a six-fold increase in investment from US$22 billion in 2005 reaching US$155 billion annually by 2008 (New Energy Finance, 2009). By 2011 the market had grown to US$234 billion;

- the creation of carbon markets where the value of annual trading volumes rose to US$122 billion by 2009. Studies estimate that emissions were reduced by around 120–300 megatonnes in the first three years of the European Union Emissions Trading System (May and Ellerton, 2008);
- the prospect of new multi-trillion dollar markets by 2020 associated with more effective management of natural resources, the provision of integrated urban environmental infrastructure and low carbon transport systems for cities, and low carbon industrial, commercial and residential real estate;
- the early maturing of micro-finance, micro-credit and micro-insurance as both forms of financial intermediation to serve those people at the base of the pyramid (Prahalad, 2005) and as an asset class in their own right.

Notably, it is estimated (IEA, 2009) that over the next 30 years US$1 trillion annually is required for the world's energy infrastructure both to maintain and extend the supply of power to more people (US$500 billion) and to finance the transition to a low carbon, cleaner energy infrastructure (US$500 billion). The projected annual shortfall to drive this low-carbon transition in the developing economies is US$350 billion. It is estimated (UNFCCC, 2007) that private sources will account for 86 per cent of the finance required for a low carbon economic transition.

Clearly, stable and resilient capital markets, supported by productive processes of investment and financial intermediation, will have a pivotal role in the provision of capital at sufficient scale for the delivery of a green economy. Early analysis would suggest that the interaction of public and private finance will be a key determinant in any such transition, particularly the extent to which smart deployment of public funds supported by a coherent policy framework, catalyses and leverages greater private investment in a green economy. Of the post crisis government stimulus packages, some US$470 billion out of US$3 trillion-plus in public funds committed[4] to head off prospects of crippling global depression was ear-marked for low carbon and environmental infrastructure investments. This massive injection of public green financing has created, in effect, a real-time 'investment laboratory' which will help to deepen our understanding of the greening of capital and finance, especially the leverage effect that the deployment of green public funds has on private investments. Within this, the role of multilateral financial institutions (MFIs), such as The World Bank, the International Finance Corporation (IFC), the 30-plus regional MFIs, as well as export credit and investment guarantee agencies, will be critical in the early formation of new markets as private finance and investment adjust to and gain confidence in evolving green economy policy frameworks.

Hurdles for a green economy:

The daunting political, economic, social and technical challenges that exist, many exacerbated by strong vested interests, would appear to weigh against a rapid transformation of capital markets and financial services in a manner that would serve a green economy. The changes required, from improving the way externalities are accounted for and priced by the markets to the reallocation of capital at scale, face many hurdles.

Furthermore, such changes in our capital market and financial services systems would appear distant if current mainstream values and beliefs – notably those prevalent ahead of the 2007–8 financial collapse in some quarters of the capital market community – continue to dominate. Also, great swathes of the modern global financial system, notably trade and commodities finance, as well as foreign exchange and money markets, have few if any links to the green economy agenda *per se* at present, and the opportunities to promote the agenda in these important facets of finance are poorly understood. Separately, the negative impacts of the 'shadow side of finance' (see Box 12.2) in terms of threats to the stability of the global financial system, and the potential impacts of such inherent instability for the delivery of a green economy, deserve examination also.

Box 12.2 **Understanding the shadow side of finance**

The 2007–2008 subprime crisis, the ensuing credit crunch and public sector responses to the severe global economic downturn have cost governments worldwide over US$10 trillion as part of their economic stimulus packages engineered to avoid a global depression and as support for failing or troubled financial institutions. Developed countries provided US$9.2 trillion and emerging economies spending reached US$1.6 trillion following the crisis. This rapid accumulation of public debt brought the G20 group a budget deficit of 10.2 per cent of economic output or GDP in 2009, the largest for most countries since World War II. Trading on over-the-counter markets, or OTC markets as they are commonly known, has come under forensic examination by lawmakers across the world concerned that these markets enabled incalculable amounts of risk to build up in the global financial system. The growth of more opaque markets, many associated with a new breed of electronic trading platforms, have also attracted what have become known as 'dark pools' of capital or liquidity that seek to benefit from the more lightly regulated markets and enable trading in equities and other instruments to be masked. Within this so-called 'shadow side' of finance, the trading and speculative roles played by, for example, hedge funds and the proprietary trading arms of banks, dealing for themselves or on behalf of large institutional investors, has come under particular regulatory examination since the financial crash. OTC markets started on a trial basis in the US in June 1990 as one of a number of market reforms to boost the transparency of stock trading in companies not listed on public exchanges. The difference between OTC markets and public markets is that on the OTC markets, where there has been an exponential increase since 2000 in the trading of a range of derivatives contracts, the quoted entities do not have listing standards. Investment banks played a pivotal role in the rapid

continued...

Box 12.2 continued

growth and development of OTC markets from the late 1990s onwards. Since the early experiments, the size of derivatives trading on OTC markets, essentially alternative trading platforms used by private counter parties to trade without the same reporting or disclosure requirements of exchange-based markets such as the New York or London Stock Exchanges, grew to US$100 trillion by 2000 and then up six-fold to US$600 trillion by late 2008, a figure equivalent to 10 times global gross domestic product. According to the International Swaps and Derivatives Association (ISDA, 2009), the outstanding notional amount of derivatives stood at over US$454 trillion in mid-year 2009, which is 30 times gross domestic product of the US. When the net figure of overlapping contracts – simply put this is contracts that cancel each other out – is calculated, the Bank for International Settlements (BIS) estimates the net figure for derivatives contracts stood at US$34 trillion by the end of 2008. The implications of the growth of OTC markets can be appreciated when it is understood that 'trading activity on the OTC is US$60 trillion dollars annually, while turnover on the public market is US$5 trillion'. Some market observers have questioned whether a global market that is only '8 per cent transparent' is actually a market. In 2010 questions are being raised over many different aspects of the governance, oversight, and practices of the investment chain and the intermediaries in the financial system, notably the investment banking sector, following the crisis. Fears over the inherent instability of the system are being raised also by institutional investors that are required to protect and grow assets over the long term. The underlying stability of the financial system and the future governance of the system have important implications for the ultimate delivery of a green economy.

It is clear that across lending, investment and insurance – the core activities of our financial service and capital market systems – significant changes in philosophy, culture, strategy and approach, notably reigning-in the overwhelming dominance of short-termism, will be required if capital and finance is to be reallocated to accelerate the emergence of a green economy. At the same time, fundamental aspects of international accounting systems and capital market disciplines, as well as our understanding of fiduciary responsibility in investment policy-making and investment decision-making, will need to evolve to fully integrate a broader range of environmental, social and governance (ESG) factors than takes place at present. Without these changes the pricing signals that would support the transition to a green economy will remain weak.

The changes, challenges and opportunities touched on briefly above frame just a few of the many issues facing finance and investment that will determine also, to a significant degree, the nature and timing of the transformation to a green economy. Such a transition will depend greatly on how the world's financial assets are managed and deployed and, as part of this, how our capital markets and financial service institutions reorient their dynamics, disciplines and operations to fully price and account for a range of externalities such as climate change, ecosystems destruction, and the effective management of waste in all its forms, and, also, importantly, how such institutions position to compete in new lucrative green markets.

The governance of the financial system and the institutions that serve it will be under a bright spotlight, intensified after the crisis of 2007–2010, as the world seeks a meaningful transition to a true green economy.

Greening signals for the investment chain

The fact that the world's largest institutional investors, including more than 900 organizations[5] representing more than US$21 trillion in assets which back the Principles for Responsible Investment (PRI),[6] have become engaged in these green economy issues sends significant signals to the rest of the investment chain (see Box 12.3). As institutional investors are considered to head the investment 'food chain', their interests and concerns are followed closely by influential pension consultants, asset management organizations, the financial analysts' community and a range of other service providers embedded in the investment chain. In summary, the management and deployment of institutional investment funds in coming decades, the majority of which are currently concentrated in developed economy savings structures, will be a pivotal factor in the emergence of a green economy, and critically, the pace at which such a transformation takes place. The need for these funds, many held in OECD countries, to benefit in their long-term investment returns from the dynamic economic growth, on-going industrialization, and access to as well as management of natural assets in the developing economies of the world, also creates additional pressures for investment institutions to consider a broader range of ESG considerations as they explore new markets outside their traditional geographic investment scope.

Furthermore, as the economic foundations of a globalizing market evolve, we are starting to see important changes in the investment industry that will accelerate in coming decades, again, with profound implications for the transition to a green economy. For example, the emergence of sovereign wealth funds over the past 20 years as well as the increasing relative importance of investment approaches such as hedge funds and the private equity sector since the mid-1990s, are indicators of fundamental changes taking place. Additionally and as noted, the role of the 8.6 million rich people – known as high-net-worth individuals and ultra-high-net-worth individuals in investment jargon – that control over US$32 trillion in assets today will also be influential in the emergence of a green economy. Research (UNEP FI Asset Management Working Group, 2007) shows that the next generation of this global community is more receptive to green economy-supporting investments around themes such as climate change, water and forestry. The potential role of the high-net-worth community in fostering a green economy is reinforced by the fact that the next 15 years will see the largest ever transfer of private wealth to a younger generation that should be more open to the idea of a green economy.

Box 12.3 **The case of the Norwegian Government Pension Fund – Global**

The Norwegian Government Pension Fund – Global, one of the largest sovereign wealth funds in the world – currently has a broad ownership in almost 8,000 companies worldwide. The fund is largely passively invested and holds an average ownership share of one per cent in each company it is invested in. As a universal owner, the pension fund believes that it will benefit from making sure that good corporate governance and environmental and social issues are duly taken into account. Having been entrusted to manage the wealth of its end-beneficiaries, fiduciary responsibility for the pension fund also means taking widely shared ethical values into account.[7] For the pension fund, ESG issues present regulatory, market, reputational, and operational risks and opportunities its shareholders need to consider in order to fully understand the companies in which their capital is invested. Hence, it aims to define robust strategies for the integration of these issues across all our investments at both a strategic and a portfolio level. Its sustainability strategy is based on three pillars:[8]

1. Research project – understanding the impact of ESG risks on wealth creation

The Norwegian Ministry of Finance, acting as principal for the fund, currently participates in a research project between the investment consultancy Mercer and 12 large international pension funds from Europe, North America and Asia. Through this research consortium the Norwegian Government Pension Fund aims to assess how the challenges of climate change may affect the financial markets and how it ought to invest in light of the fund's vulnerability to climate risks.

2. New responsible investment programme – targeting underlying portfolios that take account of environmental impacts

The Norwegian Finance Ministry is in the process of establishing a new investment programme for the fund which will focus on environmental investment opportunities, such as climate-friendly energy, improving energy efficiency, carbon capture and storage, water technology and the management of waste and pollution. The investments will have a clear financial objective. The principal is looking at several possible investment opportunities, such as green bonds issued by the World Bank. It is also looking at listed equities and overweighting companies with a good environmental profile using an index where the weight ascribed to the companies is affected by environmental criteria defined by the index provider. Initially, the Ministry aimed at investing NOK4 billion on the basis of environmental criteria in 2010.

3. Dialogue with companies

The pension fund's manager, Norges Bank Investment Management (NBIM), has set out its expectations on companies' climate change management. In its capacity as an investor, the pension fund is able to evaluate the degree to which a specific company is exposed to the risks and opportunities that arise from climate change, both in its direct operations and its supply chain. NBIM is required to consider companies' efficient adaptation to this transition with the purpose to protect the financial assets of the fund. NBIM expects companies to develop a well-defined climate change strategy. Similarly NBIM has outlined a set of expectations for corporate performance on sustainable water management.

As referenced earlier, the financial and economic crisis of 2007–2009 has prompted deep concerns within the financial policy-making community and the long-term institutional investment industry with respect to the economic, social and political risks posed by inherent instability in the global financial system being exacerbated by aggressive financial engineering and innovation centred around new products fashioned to benefit from short-term opportunities[9].

As well as these deepening concerns around how governance of the financial system and governance of financial institutions that are 'too big to fail' might undermine stability of the overall market system, there has been a gradual appreciation in parts of the investment industry over the past 20 years, that changes in public policy, over time and mainly within the developed economies, tend to drive a range of externalities closer to market relevance in that they become financially material. Indeed, in recent years, there has been a growing amount of research and evidence articulating the materiality of ESG issues to the performance of investment portfolios across companies, sectors, regions, asset classes and through time. As ESG issues become financially material through the emergence of new norms ('soft law') or in a given legal jurisdiction through policy and legislation ('hard law'). This requires an adjustment of the risk–reward equation by the investment industry both in respect of fulfilling fiduciary responsibilities to give appropriate consideration to a full range of material risks and also to be open to new investment opportunities and emerging markets created by public policy and regulatory changes. For a range of reasons, from the ideological to the technical, there has been a reluctance to see ESG integrated into investment policy and decision-making although strong policy, accounting, financial materiality and legal cases are building to drive this change within the market over time.[10]

Sovereign wealth funds

Sovereign wealth funds (SWFs) are state-owned investment funds that play an important role in financing international activities. Their assets include vast holdings of bonds, stocks and foreign currency reserves. Of late they have played a key role in the discussions of stabilizing national currencies.

Guiding their activity are the Generally Accepted Principles and Practices (GAPP) known as the Santiago Principles. These principles could be amended to direct those funds to sustainable development activities, and fiduciary responsibilities could be reformed to realign the systemic process driving kind of short-term investment thinking that helped undermine the global banking system. By shifting fund managers' incentives away from short-term gains through integrating these long-term principles and frameworks into the market system, sustainability can be built into the financial architecture which drives investment decisions. There are a number of places the principles could be amended to enable that:

GAPP 10. PRINCIPLE

The accountability framework for the SWF's operations should be clearly defined in the relevant legislation, charter, other constitutive documents or management agreement.

GAPP 11. PRINCIPLE

An annual report and accompanying financial statements on the SWF's operations and performance should be prepared in a timely fashion and in accordance with recognized international or national accounting standards in a consistent manner.

GAPP 18. PRINCIPLE

The SWF's investment policy should be clear and consistent with its defined objectives, risk tolerance, and investment strategy, as set by the owner or the governing body(ies), and be based on sound portfolio management principles.

GAPP 18.1 Sub-principle. The investment policy should guide the SWF's financial risk exposures and the possible use of leverage.

GAPP 19. PRINCIPLE

The SWF's investment decisions should aim to maximize risk-adjusted financial returns in a manner consistent with its investment policy, and based on economic and financial grounds.

GAPP 19.1 Sub-principle. If investment decisions are subject to other than economic and financial considerations; these should be clearly stated.

(IWG, 2008)

SWFs are growing in the developing world with ones in China, Brazil, Botswana, Malaysia, Chile and Algeria in the top 40. The large amounts of money could be a huge trigger for investment in sustainable development.

Credit-rating agencies

Credit agencies are private companies that assign ratings of credit worthiness for issuers of certain types of debt obligations. The most well-known are Fitch, Moody's, and Standard and Poor's. A company's or government's rating that is lowered below AAA can be extraordinarily costly in its expenses for subsequent borrowing. The most recent example was the stunning (at least to the US) down-rating of the US top-tier AAA credit rating by Standard and Poor's. Greece was only given a B rating by Fitch even after massive restructuring – which amounts to junk bond status, and therefore a non-investment grade.

In reaction to such steps, governments, particularly in the EU, have begun to propose strict reporting standards to govern those agencies. The same type of regulations could be applied to encourage sustainability. The credit rating system could play a very significant role by integrating sustainability criteria in line with the principles of the green and fair economy. A company's credit rating, and therefore its ability to borrow, would in part be made up by sustainability criteria which could include whether you produced a sustainability report for example. UNEP Finance Initiative has started work on credit-rating methods, which should be supported and accelerated.

New economic indices

Forty years ago, a passionately campaigning Robert Kennedy deconstructed the myth of linear growth:

> We seem to have surrendered community excellence and community values in the mere accumulation of material things. Our gross national product … counts air pollution and cigarette advertising, and ambulances to clear our highways of carnage … It counts the destruction of our redwoods and the loss of our natural wonder in chaotic sprawl. It counts napalm and the cost of a nuclear warhead, and armored cars for police who fight riots in our streets. It counts Whitman's rifle and Speck's knife, and the television programs which glorify violence in order to sell toys to our children.
>
> Yet the gross national product does not allow for the health of our children, the quality of their education, or the joy of their play. It does not include the beauty of our poetry or the strength of our marriages; the intelligence of our public debate or the integrity of our public officials. It measures neither our wit nor our courage; neither our wisdom nor our learning; neither our compassion nor our devotion to our country; it measures everything, in short, except that which makes life worthwhile.
>
> (Kennedy, 1968b)

The quote still seems stunningly prescient. It has taken until recently to really question the use of GDP. The current reliance on economic growth and GDP as an indicator of success has not delivered fair levels of well-being for society or individuals, and it has led to perverse outcomes. Instead we need to reassess our common values, making decisions that lead to a green and fair economy around what we really value. There is an important distinction between assessing current well-being and assessing sustainability.

The UN Statistical Commission has a working group of statisticians from all regions, developing a System of Environmental-Economic Accounts (SEEA). This is an ambitious attempt to create standardizing environmental data.

In July 2011, the United Nations' General Assembly called on its member states to recognize the importance of well-being to development, and is asking governments to review their measurement systems with a view to using well-being information to guide public policy (UN, 2011).

Another measure used is that of the ecological footprint. The approach of the ecological footprint is very similar to financial accounting:

> [It] tracks revenues and expenditures. Accounts have typically two sides. Just like financial balance sheets include both 'income' and 'expenditure'. Footprint accounts compare the availability of bio capital against the demand on bio capital.
>
> (Seaford et al., 2011)

The ecological footprint approach looks at what people consume and where they consume it. It has been estimated that 80 per cent of the world's population lives in countries where the population consume more than they domestically produce (Seaford et al., 2011).

The move to indices that can give us the information for decision-makers to ensure we start urgently moving towards a balance within the planetary boundaries is needed. How we measure and what we measure will have profound impacts on what way we live in the future.

National wealth accounting

The World Bank and the UN have undertaken a pilot scheme of comprehensive wealth accounting. It has been focused on three asset areas:

- producing capital;
- natural capital; and
- human and social capital.

The idea here is to expand national accounts so that they are more useful to policy makers. This would include more detailed statistics and the implementation of full environmental accounting. The move to focus on natural capital is particularly important for low income countries. The World Bank estimates that low income countries are resource rich in natural capital between 36 per cent to 50 per cent wealth. The pilot in ten countries will develop:

- core environmental annual accounts;
- expand ecosystem accounts; and
- accounts will be based on physical accounts for land, spatially explicit, monetary accounts and distribution of benefits to different stakeholder groups.

(Lange, 2011)

Conclusion

The crisis caused in 2008 was predicated by surprising numbers of insiders who did or should have known better, but convinced themselves that somehow they had changed the rules of capitalism. Others knew the problems but decided to keep quiet and hope to profit enough while they could; and many of us had no idea it would happen as we had put our trust in the financial institutions and our political leaders. So have we learnt the lessons?

As Matt Taibbi wrote in the February 2011 issue of *Rolling Stone* magazine:

> Not a single executive who ran the companies that cooked up and cashed in on the phony financial boom – an industry wide scam that involved the mass sale of mismarked, fraudulent mortgage-backed securities – has ever been convicted. Their names by now are familiar to even the most casual Middle American news consumer: companies like AIG, Goldman Sachs, Lehman Brothers, JP Morgan Chase, Bank of America and Morgan Stanley. Most of these firms were directly involved in elaborate fraud and theft. Lehman Brothers hid billions in loans from its investors. Bank of America lied about billions in bonuses. Goldman Sachs failed to tell clients how it put together the born-to-lose toxic mortgage deals it was selling. What's more, many of these companies had corporate chieftains whose actions cost investors billions – from AIG derivatives chief Joe Cassano, who assured investors they would not lose even 'one dollar' just months before his unit imploded, to the $263 million in compensation that former Lehman chief Dick 'The Gorilla' Fuld conveniently failed to disclose. Yet not one of them has faced time behind bars.

(Taibbi, 2011)

This cannot be allowed to happen again, and certainly not with our planet's all-too-finite natural resources. Millions of people in developed countries have already suffered because of the actions of a relatively small number of individuals. Further tens of millions of people in developing countries may suffer in coming years from drastically lowered availability of funds for ODA and direct investment. If the world is not able to move to an economy based on sustainability criteria, it is all of us who will suffer the consequences.

To do so will require concerted change in our consumption and production patterns. It will require that we regulate to ensure that companies operate within planetary boundaries. It will change the way that companies themselves are valued, and their balance sheets in the future must take account of environmental externalities. The work of the finance industry must be rigorously monitored as well.

A series of creative and powerful proposals have emerged in the area of sovereign wealth funds: Earth Bonds and Green Bonds which would attract private funds that could contribute immensely to helping fund the development and application of green technologies and the establishment of a global green economy. In addition to sovereign and private funds, ODA will need to play an increasing role to help facilitate the viable development of LDCs.

One of the central challenges governments face as they move into the twenty-first century is ensuring that developing countries can meet their rapidly growing energy needs in ways that will not move the human community beyond the thresholds of the planet's limits. We cannot expect them to respond to mere exhortations to not repeat the wasteful and harmful practices by which the older industrialized countries have moved dangerously close to these thresholds. This is the explicit premise of the UN Framework Convention on Climate Change (UNFCCC), which places the burden on OECD countries, which need to set an example by effecting major efficiencies in their own use of energy and leaving space for them to grow.

They need to also set an example in corporate responsibility and targeting of green investing, and must ensure that developing countries have access to the latest, most environmentally sound and efficient energy technologies and the finances they require to afford them. They must take the lead in research and development of new technologies and finding solutions to the many unresolved energy challenges – if not within the UNFCCC, then with governments taking active responsibility for defining and adhering to alternatives.

We are convinced that – given the long lead times that characterize major changes in the energy industry, the enormity of the problems in the finance sector, and the extraordinarily broad constellation of solutions that will be required to reform consumption and production patterns, food and agricultural systems, urban development, and the distribution of water and the unsustainable use of our oceans and forest resources – the decisions that shape our collective future are those that must be taken today. By the time the problems that will arise from our inaction have become acute, it will be too late to fix them.

We are the first generation in history to have the ability and responsibility to determine the future of life on Earth. We cannot afford to be complacent in the belief that whatever we do, life will go on. We must realize that the conditions which make life possible as we know it have only existed for a very brief period of our planet's long history and within very narrow limits. It is clear that humans are now impinging on these limits at a speed and on a scale beyond our ability to adjust or adapt. Human existence is at real and imminent risk. But the pathways to success, however challenging, are also very real.

The focus on green economy, or how we make all economies take sustainable development concerns to their centre, is the defining question of our time. The faster we can start to address it, the longer we will have to get it right and face other problems – and the greater the chance to fairly and equitably share the resources on this planet.

Notes

1 BRICs – acronym used by Jim O'Neill of Goldman Sachs from 2001 onwards. The four countries, combined, currently account for more than 25 per cent of the world's land area and in excess of 40 per cent of world population.

2 For example: the 'Black Monday' market crash of 1987; the Mexican 'Peso' Crisis 1994; the Asian Crisis 1997; the Russian 'Rouble' Crisis 1998; the collapse of US hedge fund Long-Term Capital Management (LTCM) 1998; the Dot.Com 'Boom and Bust' of 1999–2000; and the market collapse catalysed by corporate governance failures (e.g. ENRON, Worldcom, Parmalat) of 2001–2002.

3 The insurance industry alone had turnover of US$1.7 trillion in 2008.

4 HSBC (2009) estimated that US$470 billion of government funds were allocated to climate change themed investment as part of the global stimulus packages. The insurance industry alone had turnover of US$3.7 trillion in 2008.

5 Figures presented to the PRI Board Meeting, Offices of Investec, Durban, South Africa, December 2011.

6 The Principles for Responsible Investment (www.unpri.org) is a voluntary initiative launched by former UN Secretary-General Kofi Annan in April 2006. The PRI was incubated by UNEP FI and the UN Global Compact during 2003–2006.

7 Fiduciary responsibility – legal and practical aspects of integrating environmental, social and governance issues into institutional investment, UNEP FI, June 2009.

8 'A legal framework for the integration of environmental, social and governance issues into institutional investment', Freshfields Bruckhaus Deringer and UNEP FI, October 2005.

9 UNEP FI Asset Management working group (AMwg).

10 The 'Materiality Series' comprises three reports published by the UNEP Finance Initiative Asset Management working group (Mat I: 'The Materiality of Social, Environmental and Governance Issues to Equity Pricing', June 2004; Mat II: 'Show Me the Money: Linking ESG Issues to Corporate Value', July, 2006; Mat III 'The Materiality of Climate Change: How Finance Copes with the Ticking Clock', October 2009).

Chapter 13

A survival agenda

Twenty-one actions to help save the planet

The environmental challenges of the next few years will be intense, the economic risks will be severe, and the diplomatic and political obstacles will be formidable.

There are strategic pathways that can be taken that avoid the looming abyss of climate and ecosystem collapse, that can lead us along the ledge of equitable economic recovery, and carry us safely down into the valley of cooperation and sustainability.

Below we highlight 21 actions that could make a significant difference in determining what kind of planet human beings live on in 2032 – 20 years in the future from when this book is being written. We call this 'the survival agenda', not because life on the planet will otherwise end, but because what we collectively do today will irrevocably determine how healthily, how equitably, and how sustainably we are all able to live on the planet in an all-too-rapidly approaching tomorrow.

We suggest these with a realistic understanding of the limits of what governments feel they are able to do, but also with an equally-realistic optimism – based on historical and political experience – that at certain moments in time, when the common vision is raised, when the consequences are clear, and when the crises are imminent, we human beings – acting as governments, organizations, stakeholders, and individuals – can lift ourselves to take creative leaps that confront and overcome the most overwhelming obstacles.

The defining questions, as always, are do we care enough, do we trust enough, and do we have the courage to love enough to work together to do so?

The survival agenda

1 Adopt the Earth Charter

A declaration of fundamental ethical principles for building a just, sustainable and peaceful global society in the twenty-first century, the Earth Charter seeks to inspire in all people a new sense of global interdependence and shared responsibility for the well-being of the whole human family, the greater community of life, and future generations. It is a vision of hope and a call to action.

The Charter recognizes that ecological protection, the eradication of poverty, equitable economic development, respect for human rights, democracy, and peace are interdependent and indivisible. It provides an inclusive, integrated ethical framework to guide the transition to sustainable ways of living and a sustainable future.

2 Agree a new global climate agreement

Agree a new global climate agreement that reduces greenhouse gases so that the temperature rise is within 1.5°C.

3 Strengthen the United Nations

Strengthen the United Nations through appropriate funding and political mandate to address the challenges of the future.

4 Adopt a rights-based approach

There is a need to propose an explicit global social contract, instead of dealing with social issues as a 'safeguard' type of concern. A true rights-based approach to dealing with welfare, well-being and environmental issues is essential to sustainable development. Such an approach would put people at the heart of development that is also sustainable.

5 Acknowledge environmental limits

There is an urgent need to formally recognize key environmental limits and processes within which we must remain, and the thresholds that we must respect in order to maintain the sustainability of our planet.

6 Implement sustainable management of natural resources and capitals

The poor management and regulation of natural assets and ecosystems lead to increasingly frequent and severe regional and global crises, and is a major factor behind food, water and energy insecurity. These issues are highlighted in Agenda 21, chapters 9–18 in particular. All levels of government must ensure that development strategies take full account of the state of natural assets and ecosystems and their role in sustaining human and animal well-being and economic activity; actively investing in their conservation and enhancement to avoid a devastating and irreversible global crisis.

7 Move beyond GDP

The current reliance on economic growth and GDP as an indicator of success has led to perverse outcomes. It has not delivered fair levels of well-being for society or individuals. GDP is an inadequate metric through which to gauge well-being over time. Instead we need to reassess our common values, making decisions that lead to a green and fair economy around what we really value. There is an important distinction between assessing current well-being and assessing sustainability.

8 Reform the international financial institutions

As discussed in Agenda 21, chapters 33 and 38, there must be better incorporation of sustainable development parameters in the existing international financial institutions (IFIs), particularly in terms of funding, operations, strategic plans, objectives and implementation. Additionally, governments should seek to strengthen the efficiency of the Global Environment Facility (GEF).

9 Agree Sustainable Development Goals

In August 2011, the government of Colombia proposed negotiating Sustainable Development Goals (SDGs) as a foundation for building international political commitment at Rio. Providing measurable, tangible goals, the SDGs would address the Agenda 21 aims produced at Rio 20 years ago. The SDGs would apply in all countries, and therefore act as a complementary, successor framework to the Millennium Development Goals (MDGs), which are scheduled to conclude in 2015 and focus mainly on the global South. SDGs would also shift the centre of gravity away from the dominant economic (poverty reduction) pillar of the MDGs and more towards the environmental and social pillars of sustainable development, for example, by providing measurements against metrics of planetary boundaries, and maintaining a strong focus on consumption patterns in the global North. However, the SDGs should not detract from the urgent need for a post-2015 framework that focuses on poverty or from funding that agenda.

10 Improve international cooperation and development aid

As outlined in Agenda 21, chapter 33, future agreements concerning sustainable development financing should be centred around measureable and time-bound targets, to ensure the finance committed to implementing future targets to developing countries is truly delivered. Improving the quality of aid and ensuring it is delivered on the ground is as important as increasing the amount of that assistance.

11 Empower national, local and regional governance

National Sustainable Development Strategies should be revived and refreshed with full engagement and support from business and all parts of civil society. These strategies should be underpinned with route maps outlying national actions towards a green and fair economy. Advisory bodies such as councils for sustainable development need to be adequately resourced to play their full part in bringing forward new thinking and maintaining pressure for progress.

12 Appoint a High Commissioner or Ombudsperson for Future Generations

The needs of future generations are a crucial element of sustainable development, but are not represented in the relevant decision-making processes. A way to remedy this situation and ensure that long-term interests are heeded would be to create a High Commissioner/Ombudsperson for Future Generations at the UN and national levels. Such an ombudsperson would be responsible for and answer to youth and the not-yet-born children of all nations – and not to political officials.

13 Establish an International Court for the Environment

Environmental problems extend across international boundaries, but there are few effective international institutions to deal with them properly. Strengthening international environmental law mechanisms are essential to securing sustainable development. For example, this could take the form of an International Court for the Environment, which would build trust, harmonize and complement existing legal regimes and provide clarity and access to justice as well as redress.

14 Negotiate a global convention on Rio Declaration Principle 10

Access to environmental information, participation in transparent decision-making processes, and access to judicial and administrative proceedings should be basic rights for all, at all levels of decision-making, including local, national and international processes. Worldwide implementation of Principle 10 of the 1992 Rio Declaration is a priority. This could take the form of regional replications of the Aarhus Convention in other parts of the world, or even more widespread adoption of the Aarhus Convention.

15 Approve a Convention on Corporate Social Responsibility and Accountability

All corporations should be mandated to report on their environmental impacts and contribution to well-being, or explain why they are not doing so. There should be a commitment to develop national regulations which mandate the integration of material sustainability issues in the annual report and accounts; and therefore providing effective mechanisms for investors to hold companies to account on the quality of their disclosures. A corporate accountability convention should establish social and environmental duties for corporations and ensure the international direct liability of corporations.

16 Decide that all subsidies that undermine sustainable development should be eliminated

Particularly subsidies underpinning fossil fuel use and unsustainable agricultural and fishery practices.

17 Begin comprehensive, full-cost fiscal reform

True environmental costs of production and consumption must be internalized into accounting models in order to address causes rather than simply the symptoms of environmental degradation. The polluter-pays principle should be adopted in practice in standard accounting and reporting practices for both business and governments, so that these costs can be reflected in market valuations and environmental impact assessments. Furthermore, green taxes should be used to incentivize positive behaviors and discourage harmful ones.

18 Set up a global Green Bond investment system

Set up a global Green Bond investment system under the IFC, to help fund the transition to a green economy.

19 Impose a global Financial Transaction Tax (FTT)

Impose a global Financial Transaction Tax on all currency exchanges and speculative trades in shares, with a significant proportion of the revenue used to support long-term efforts to fight climate change in developing countries and implement sustainability programmes.

20 Amend the Santiago Principles

Amend the Santiago Principles that guide Sovereign Wealth Funds to include sustainable development among their core goals and principles.

21 Restructure credit-rating agencies

Restructure credit-rating systems to include sustainable development criteria in the financial instruments of all governments.

Chapter 14

Which road now?

The question posed repeatedly throughout this book, explicitly and implicitly, is 'Will human beings gain the awareness, the capacity and the courage to significantly modify the path of their economic and social behavior, and pioneer a less familiar but more creative route that can allow them to co-exist safely within their world's ecological limits?'

And will they decide to do so before the gathering storms of environmental, economic and social crises render such transitions impractical, unachievable or simply insufficient?

The decisions being taken by the world's governments at the 2012 Summit will be a significant factor in determining the answer to those questions.

But more significant will be the actions of those governments – and the actions of tens of thousands of individual consumers and producers, business owners and workers, teachers and students, parents and children – *after* Rio+20.

For the roll-call vote that will determine the success or failure of sustainable development – and the survivability of modern cultures – will be the sum total of all the actions of all the members of all our countries for the next twenty years.

That 'vote' can move towards embracing the myriad of sustainable policies and technologies that have emerged in the past few decades – with those that existed for millennia before being almost entirely forgotten.

Those actions and decisions, however, are being taken today in a global political environment of unusually widespread confusion and stress. Developed countries' economies are still teetering at a fragile non-equilibrium as a result of the global financial industry's failure and fraud. Developing countries are dealing with not only reduced flows of private trade and investment, but – yet another – rationale for promises not being fulfilled on provision of technology transfer and ODA.

Partly because of that economic instability, deep political discontent and division – expressed in some places with astonishing courage and altruism, and in others with dangerous cynicism and greed – has bubbled to the surface. Democratic movements expressing the highest idealism, and nationalist movements provoking the darkest fears, are grappling in intense, asymmetrical and often dishonest hand-to-hand combat – and the universally critical results are far from certain.

Meanwhile, governments – the permanent sector that should be helping to calm the increasing sense of chaos, and the desperation and fear of individuals – seem all too often immobilized. Not surprisingly, public confidence in government and governing institutions, in all global regions, is extraordinarily, historically low.

When human societies are faced with massive internal stress or external threat, there are generally only two broad options for action. Either they – we – all band together, and resolve to work together to share the sacrifice and face the challenge cooperatively. Or they – we – fragment into smaller social units – usually tribes or extended families – start to accuse 'outsiders' of causing all problems, and turn against each other as panic and chaos take hold.

In an era of extensive social interdependence, intensive technological dependence, and increasing biological vulnerability, the second scenario is not a survivable option.

Ultimately the choice we face is as old and profound – and as clear and simple – as those posited by the world's great religions.

Will we succumb to fear and ignorance, in the form of nationalistic aggrandizement and personal greed, or will we rise above the temptation to panic, and work together for the common good? Will we find ways to help each other by creating economies that work in balance, or will we relentlessly strive to succeed for only ourselves, at the expense of everyone else? Will we cooperatively share the natural resources of this extraordinarily beautiful and bountiful planet, or will we instead destroy those resources, through hoarding and overuse?

Will we, in the words of one of the world's all-time great teachers, choose fear and death? Or will we choose hope and life?

The answer isn't yet clear. But the source of that answer is.

The choice is ours.

References

All websites accessed March 2012

Aalborg Commitments (2004) *Inspiring the Future*, Bonn: ICLEI.

Agawal, A. (1992) 'Nature – Man who never gave up hope'. Available online at http://www. lifepositive.com/body/nature/environmentalist.asp

Alliance for Responsible Atmospheric Policy (n.d.) Available online at http://www.arap.org/

Amil, F. (2006) 'Speech to the UN Security Council'. Available online at http://news.bbc. co.uk/2/hi/science/nature/6665205.stm

Annan, K. (1999) 'Address to the World Economic Forum', Davos, Switzerland. Available online at http://www.un.org/News/Press/docs/1999/19990201.sgsm6881.html

Annan, K. (2005) *In Larger Freedom: Towards Development, Security and Human Rights for All*, New York: United Nations.

ANPED (2001) 'NGO statement to UNCEC Regional Preparatory Meeting for WSSD'. Available online at http://www.johannesburgsummit.org/html/prep_process/europe_ northamerica/final_unece_ngo_statement.pdf

Armstrong, N. (1969) 'Speech from the moon'. Available online at http://en.wikiquote.org/ wiki/Neil_Armstrong

Asmal, K. (2000) 'First World chaos, Third World calm: A multi stakeholder process to part the waters in the debate over dams', *Le Monde*, 15 November.

Auken, S. (1996) 'Speech to the UN CSD', in United Nations Commission on Sustainable Development, *Three Years since the Rio Summit*, London: Stakeholder Forum, London.

Authers, J. (2010) 'Market forces', *Financial Times*, 22 May.

AVIVA (2011) *Earth Summit 2012: Towards a Convention on Corporate Sustainability*. London: AVIVA.

Ban Ki-moon (2011) Secretary-General's remarks to the World Economic Forum Session on Redefining Sustainable Development. Speech at Davos, Switzerland, 28 January 2011. Available online at http://www.un.org/sg/statements/?nid=5056

Banuri, T. (2009) 'Sustainable development agreement signals new cooperation on finding solutions to global crises', UN, New York. Available online at http://www.un.org/News/ Press/docs/2009/envdev1054.doc.htm

Benson, E. (2010) *Stakeholder Empowerment Project*, London: Stakeholder Forum.

Biermann, F. (2011) *Reforming Global Environmental Governance: The Case for a United Nations Environment Organization (UNEO)*, London: Stakeholder Forum.

Biermann, F. and Bauer, S. (2004) 'Does effective international environmental governance require a world environment organization? The state of the debate prior to the report of the high-level panel on reforming the United Nations'. Global Governance Working Paper No 13. Amsterdam, Berlin, Oldenburg, Potsdam: The Global Governance Project. Available online at http://www.glogov.org/images/doc/WP13.pdf

Black, R. (2009) 'Why did Copenhagen fail to deliver a climate deal?' Available online at http://news.bbc.co.uk/1/hi/8426835.stm

Bloomberg New Energy (2012) 'Solar surge drives record clean energy investment in 2011'. Available online at http://bnef.com/PressReleases/view/180

Bloomberg New Energy Finance (2009) Bloomberg New Energy Finance, 4–6th March 2009 Summit. Available online at http://2009.newenergyfinancesummit.com/nef%20 2009_lores2.pdf

Browning, S. E. (2007) 'Exorcising Ghosts of Octobers Past', *The Wall Street Journal*, 15 October.

Brundtland, G. H. (1996) 'Speech to the Oslo Round Table on Consumption and Production Patterns', in United Nations Commission on Sustainable Development, *Three Years since the Rio Summit*, Stakeholder Forum, London.

Brundtland Bulletin (1992) *Centre for Our Common Future*. No. 1 (Sept. 1988)-issue 18 (Dec. 1992). Geneva: The Centre.

Brusasco-Mackenzie, M. (2002a) 'Environment and security', in F. Dodds (ed.) *Earth Summit 2002: A New Deal*, London: Earthscan, London.

Brusasco-Mackenzie, M. (2002b) 'The road to Johannesburg after September 11', in Kurt Klotzle (ed.) *The Road To Johannesburg After September 11, 2001*, World Summit Papers of the Heinrich Böll Foundation 9. Berlin: Heinrich Böll Foundation, Berlin. Available online at www.worldsummit2002.org/publications/WSP9.pdf

Bush, G. (1992) President George Bush, Sr. at the 1992 Earth Summit. Rio de Janiero, 1992.

Cardoso, F. H. (2004) *Report of the Panel of Eminent Persons on United Nations–Civil Society Relations*, New York: United Nations.

Cavalcanti, H. (1995) 'Speech to the UN Commission on Sustainable Development', in United Nations Commission on Sustainable Development, *Three Years since the Rio Summit*, London: Stakeholder Forum.

CBD (2010) 'A new era of living in harmony with Nature is born at the Nagoya Biodiversity Summit', UN Convention on Biological Diversity, Montreal. Available online at http://www.cbd.int/doc/press/2010/pr-2010-10-29-cop-10-en.pdf

Centre for Our Common Future (1990) *Centre for Our Common Future*, no. 1, July 1990. Geneva: Centre for Our Common Future.

Charnovitz, S. (2002) 'A World Environment Organization', *Columbia Journal of Environmental Law* 27(2): 321–357. Available online at http://www.wilmerhale. com/files/Publication/6ad8618d-6535-4a81-8046-cd084016b0f2/Presentation/ PublicationAttachment/d61941b0-fa16-4ad1-9e19-254449cf83be/Charnovitz1.pdf

Chirac, J. (2002) Statement to the World Summit on Sustainable Development, Johannesburg, 2 September. Available online at http:// www.un.org/events/wssd/statements/franceE. htm

Chirac, J. (2003) Speech at the opening of the fifty-eighth session of the United Nations General Assembly, New York, 29 September.

Clinton, B. (1997) Speech to UN General Assembly Special Session on Rio, United States, 16 November. Available at daccess-ddsny.un.org/doc/UNDOC/PRO/N97/857/61/PDF/ N9785761.pdf?OpenElement

Clinton, B. (1999) Presidential Executive Order 13141 – Environmental Review of Trade Agreements. Available online at http://www.presidency.ucsb.edu/ws/index. php?pid=56947

Collett, S. (1990) 'Memo to the NGO Development Committee', Quakers Office to the UN. Posted on the electronic conference.

Colombia, government of (2011) RIO + 20: Sustainable Development Goals (SDGs), An updated Proposal from the Governments of Colombia, Guatemala, Bogota, 2011.

CPEG (2010) *Taxing Main Street to Revive Main Street*, Chicago, IL: Chicago Political Economic Group.

Daley, H. (1977) *Steady-State Economics*, San Francisco, CA: WH Freeman.

da Silva, L. (2007) Speech to the UN General Assembly, United Nations, New York, 25 September. Available online at http://www.un.org/apps/news/story.asp?NewsID=2395 2&Cr=general&Cr1=debate

Davies, Z., Edmondson, J. and Herinem, A. (2011) 'Mapping an urban ecosystem service: quantifying above-ground carbon storage at a city-wide scale', *Journal of Applied Ecology*, 48(5): 1125–34.

de Boer, Y. (2006) 'Looking ahead to CSD15 – an interview with Yvo de Boer', *Outreach*, 12 May. London: Stakeholder Forum. Available online at http://www.unedforum.org/news/outreach/csd14/Friday12May.pdf

DECC (2011) 'CRC energy efficiency scheme', Department of Energy and Climate Change. Available online at http://www.decc.gov.uk/en/content/cms/emissions/crc_efficiency/crc_efficiency.aspx

Desai, N. (1995) 'Speech to the UN Commission on Sustainable Development', in United Nations Commission on Sustainable Development, *Three Years since the Rio Summit*, London: Stakeholder Forum.

Desai, N. (2001) Statement by UN Under Secretary-General to the World Summit on Sustainable Development, New York, 30 April. Available online at http:// http://www.johannesburgsummit.org/html/documents/statement_by_nitin_desai_to_prepcom_one.htm

Djoghlaf, A. (2010) Statement by Ahmed Djoghlaf. Executive Secretary Convention on Biological Diversity on the occasion of the 65th Session of the United Nations General Assembly. Available online at http://www.cbd.int/doc/speech/2010/sp-2010-11-01-un-en.pdf

Dodds, F. (2000) *Earth Summit 2002 Non Paper*, London: UNED Forum.

Dodds, F. (2001) *Inter-linkages of Multilateral Environmental Agreements*, London: Stakeholder Forum.

Dodds, F. (2011) Speech at the closing of the 64th Annual United Nations Department of Public Information Non-Governmental Organizations Conference: Sustainable Societies – Responsive Citizens, 3rd–5th September.

Dodds, F., McCoy, M. and Tanner, S. (1997) 'NGO priority issues for UNGASS', in F. Dodds and M. Strausd (eds) *How to Lobby at Intergovernmental Meetings*, London: Earthscan.

Dunion, K. (1998) Speech by Chairman of Friends of the Earth International at UNCSD Multi-Stakeholder Dialogue, 11 March 1998.

ECLAC (2011) *Sustainable Development in Latin America and the Caribbean 20 Years On from the Earth Summit: Progress, Gaps and Strategic Guidelines*, Santiago: Economic Commission for Latin America and the Caribbean.

ECO (1972a) 'Eco-barbarians or new leaders?' 6 June, produced by *The Ecologist* and Friends of the Earth.

ECO (1972b) 'What makes Uncle Sammy run'. 10 June, produced by *The Ecologist* and Friends of the Earth.

ECO (1972c) 'After Stockholm, what now?'. 16 June, produced by *The Ecologist* and Friends of the Earth.

El-Ashry, M. (2011) *Sustainable Development Governance and a Sustainable Development Board*, London: Stakeholder Forum.

ELCI (1991) *Agenda ya Wananchi*, Nairobi: Environment Liaison Centre International.

European Union (2006) *European Strategy on Sustainable Development*, Brussels: European Commission.

Figueres, C. (2010) 'UN Climate Change Conference in Cancún delivers balanced package of decisions, restores faith in multilateral process', press release. Available online at http://www.unhabitat.org/downloads/docs/9280_99927_pr_20101211_cop16_closing.pdf

Friedman, T. (2006) quoted in 'Journey to Planet Earth', Public Broadcasting Corporation,. Available online at http://www.pbs.org/journeytoplanetearth/plan_b/index.html

Friends of the Earth (2002) *Briefing on Corporate Accountability*, London: Friends of the Earth.

Friends of the Earth (2009) *Over Comsumption? Our Use of the World's Natural Resources*, Brussels: Friends of the Earth Europe. Available online at. http://www.foe.co.uk/resource/reports/overconsumption.pdf

Frumin, E. (1998) Speech by International Confederation of Free Trade Unions at UNCSD Multi-Stakeholder Dialogue, 12 March 1998.

G20 (2009) G20 Finance Ministers and Central Bank Governors Communiqués, Gyeongju, Korea, October 23. Available online at http://www.g20.utoronto.ca/2010/g20finance101023.html

G77 (2008) Statement by H. E. Ambassador Byron Blake, Delegation of Antigua and Barbuda on Behalf of the Group of 77, United Nations, New York, 27 October. Available online at http://www.g77.org/statement/getstatement.php?id=081027b

GEF (2011) 'What is GEF'. Available online at http://www.thegef.org/gef/whatisgef

Goldtooth, T. (2010) www.stakeholderforum.org/sf/outreach/index.php/day2-item4

Göpel, M. (2011) *Ombudspersons for Future Generations as Sustainability Implementation*, London: Stakeholder Forum.

Gore, A. (1992) *Earth in the Balance*, London: Earthscan.

Gore, A. (1993) Keynote Address to the UN Commission on Sustainable Development, United Nations, New York, 14 June. Available online at http://clinton1.nara.gov/White_House/EOP/OVP/html/sustain.html

Gormley, J. (2008) 'Open, unscripted and interactive', interview in *Outreach*, 16 May, London: Stakeholder Forum. Available online at www.sdin-ngo.net/outreachissues/pdf/080516-outreachissues.pdf

Griffiths, H. (2005) 'Human and environmental rights: The need for corporate accountability', in F. Dodds and T. Pippard (eds) *Human Environmental Security: An Agenda for Change*, London: Earthscan.

Gummer, J. (1995) 'Speech to the UN Commission on Sustainable Development', in United Nations Commission on Sustainable Development, *Three Years since the Rio Summit*, London: Stakeholder Forum.

Hamer, M. (2002) 'Plot to undermine global pollution controls revealed', *New Scientist*, 2 January. Available online at http://www.newscientist.com/article/dn1734-plot-to-undermine-global-pollution-controls-revealed.html

Hare, B. (1997) 'Clinton gives in to oil, coal and car industry at Earth Summit', Greenpeace press release, 26 June. Available online at http://archive.greenpeace.org/comms/97/summit/prjune26.html

Havel, V. (1994) quoted in 'Vaclav Havel dies at 75: Czech leader of '89 "Velvet Revolution",' *Los Angeles Times*, 18 December 2011. Available online at http://articles.latimes.com/2011/dec/18/local/la-me-1219-vaclav-havel-20111219-5

Heinberg, R. (2011) *The End of Growth*, Gabriola Island: New Society Publishers.

Hemmati, M., Dodds, F., Enayati, J. and McHarry, J. (2002) *Multi-Stakeholder Processes for Governance and Sustainability: Beyond Deadlock and Conflict*. London: Earthscan.

Henderson, M.C. (2010) *The 21st Century Environmental Revolution*. Waves of the Future

Hohnen, P. (2002) 'One step beyond multi-stakeholder processes', in M. Hemmati, F. Dodds, J. Enayati and J. McHarry (eds) *Multi-Stakeholder Processes for Governance and Sustainability: Beyond Deadlock and Conflict*, London: Earthscan.

HSBC (2009) *Building a Green Recovery*, London: HSBC. Available online at http://www.hsbc.com/1/PA_esf-ca-app-content/content/assets/sustainability/090522_green_recovery.pdf

Hungary (1993) Act LIX of 1993 on the Parliamentary Commissioner for Civil Rights (Ombudsman). Available online at http://www.worldfuturecouncil.org/fileadmin/user_upload/papers/Ombudsman_Act.doc

Huntsman, J. (2011) 'CBS Republican Presidential Foreign Affairs debate'. Available online at http://www.cnn.com/2011/11/12/opinion/martin-gop-wall-street/index.html

IACSD (1993) Mandate of the Inter-Agency Committee on Sustainable Development, United Nations, New York. Available online at http://www.un.org/esa/documents/acc.htm

ICE Coalition (2011) *An International Court for the Environment*, London: Stakeholder Forum.

ICTSD (1999) 'No new issues without redress of Uruguay round imbalances', *ICTSD Bridges Weekly*, 3(46), 24 November.

IEA (2009) *Meeting Energy Challenges in Public–Private Partnership*, Paris: International Energy Agency (IEA).

IIED (2005) *From Profiles of Tools and Tactics for Environmental Mainstreaming*, no. 11, London: Council for Sustainable Development, IIED.

IISD and UNEP FI (2010) *Financial Stability and Systemic Risk: Lenses and Clocks*, UNEP Finance Initiative and the International Institute for Sustainable Development (IISD). Available online at http://www.iisd.org/pdf/2010/trade_lenses_and_clocks.pdf

Indonesia (2011) Chair's Summary, High Level Dialogue on Institutional Framework for Sustainable Development, Solo, Indonesia, 19–21 July. Available online at http://www.uncsd2012.org/rio20/content/documents/Chairs%20Summary%20from%20Solo%20meeting.pdf

International NGO Forum (1992a) 'Peoples Earth Declaration: A Proactive Agenda for the Future'. Available online at http://www.stakeholderforum.org/fileadmin/files/Earth_Summit_2012/People_Earth_Declaration.pdf

International NGO Forum (1992b) 'Alternative Non-Governmental Agreement on Climate Change'. Available online at http://habitat.igc.org/treaties/at-23.htm

International NGO Forum (1993) 'International NGO Forum: International Steering Committee (1992–1993)', *The Network* No. 29 (August), Geneva: Centre for Our Common Future.

International Trade Union Confederation (2011) Submission to the zero draft for Rio+20, Paris: ITUC, Paris. Available online at http://www.uncsd2012.org/rio20/index.php?page=view&type=510&nr=42&menu=20

INTGLIM (1997) *Renewing the Spirit of Rio: The CSD, Agenda 21, and Earth Summit +5*, New York: International NGO Task Group on Legal and Institutional Matters (INTGLIM). Available online at: http://habitat.igc.org/csd-97/riointro.htm

ISDA (2009) ISD Market Survey to End 2008, New York: ISDA. Available online at http://www.isda.org/statistics/pdf/ISDA-Market-Survey-annual-data.pdf

IUCN (1980) *World Conservation Strategy: Living Resource Conservation for Sustainable Development*, Gland, Switzerland: IUCN.

Ivanova, M. (2011) *Global Governance in the 21st Century: Rethinking the Environmental Pillar*, London: Stakeholder Forum.

IWG (2008) *Sovereign Wealth Funds: Generally Accepted Principles and Practice – 'Santiago Principles'*, Kuwait: International Working Group of Sovereign Wealth Funds. Available online at http://www.iwg-swf.org/pubs/gapplist.htm

Jackson, T. (2009) *Prosperity Without Growth: Economics for a Finite Planet*, London: Earthscan.

Jintao, H. (2005) Statement by President Hu Jintao of China at the High-Level Meeting on Financing for Development at the United Nations Summit. Available online at http://www.china-un.org/eng/zt/shnh60/t212916.htm

Kaniaru, D. (2007) *The Montreal Protocol: Celebrating 20 Years of Environmental Progress*, London: Cameron May Press.

Kennedy, R. (1968a) 'On the Mindless Menace of Violence', speech at the City Club of Cleveland, Cleveland, OH, 5 April. Available online at htpp://en.wikiquote.org/wiki/Robert_F._Kennedy

Kennedy, R. (1968b) Speech to University of Kansas, Lawrence, KS, 18 March. Available online at http://www.jfklibrary.org/Research/Ready-Reference/RFK-Speeches/Remarks-of-Robert-F-Kennedy-at-the-University-of-Kansas-March-18-1968.aspx

Keynes, J. M. (1936) *The General Theory of Employment, Interest and Money*, London: Macmillan.

Khor, M. (1992) 'Third World Network's analysis of the Earth Summit: Disappointment and hope', *Terra Viva*, 9 June, p. 8.

Khosla, A. (2002) 'The road from Rio to Johannesburg', *Millennium Paper* Issue 5, London: Stakeholder Forum.

Kohl, H. (1997) Speech of Dr Helmut Kohl, Chancellor of the Federal Republic of Germany, at the Special Session of the General Assembly of the United Nations, New York, 23 June.

Kompridis, N. (2009) 'Technology's challenge to democracy', *Parrhesia*, no. 8, p. 31. Available online at http://www.parrhesiajournal.org/parrhesia08/parrhesia08_kompridis.pdf

Lange, M. (2011) *The World Bank's Work of Statistics for Green Growth*, Washington, DC: World Bank.

Lerner, S. (1992) *Earth Summit: Conversations with Architects of an Ecologically Sustainable Future*, Bolinas, CA: Common Knowledge Press.

Levin, C. and Coburn, T. (2011) *Wall Street and the Financial Crisis*, Washington, DC: US Senate.

Lewis, M. (2009) 'Wall Street on the tundra', *Vanity Fair*, April.

Li, M. (2011) *Peak Energy and the Limits to Global Economic Growth*, Salt Lake City, UT: Department of Economics, University of Utah. Available online at http://www.econ.utah.edu/~mli/Annual%20Reports/Annual%20Report%202011.pdf

Lindner, C. (1992) 'The Centre for Our Common Future: Rio Reviews', *Network*, November, Geneva: Centre for Our Common Future.

Lindner, C. (1993) *The Earth Summit's Agenda for Change*, Geneva: Centre for Our Common Future, Geneva (reprinted 2005 in *The Plain Language Guide to the World Summit on Sustainable Development*, London: Earthscan).

Lindner, C. (1997) 'Agenda 21', in F. Dodds (ed.) *The Way Forward Beyond Agenda 21*, London: Earthscan.

Llewellyn, A. B., Hendrix, J. P. and Golden, K. C. (2008) *Green Jobs: A Guide to Eco-Friendly Employment*, Avon, MA: Adams Media.

May, J. L. and Ellerman, D. A. (2008) *The European Union's Emissions Trading System in Perspective*, Boston, MA: Massachusetts Institute of Technology.

Mbeki, T. (2002) 'Opening of the World Summit on Sustainable Development', Address of the President of the Republic of South Africa, Johannesburg, 26 August 2002. Available online at http://www.dfa.gov.za/docs/speeches/2002/mbek0826.htm

McConnell, F. (1994) The Biodiversity Convention: A Negotiating History, London: Kluwer Law International.

McCoy, M. and McCully, P. (1993) The Road from Rio: An NGO Action Guide to Environment and Development, The Hague: International Books.

McKinsey (2009) Global Capital Markets: Entering a New Era, New York: McKinsey Global Institute.

Meadows, D. H., Meadows, D. L., Randers, J. and Behrens III, W. W. (1972) The Limits to Growth, New York: Universe Books.

Meadows, D. L., Randers, R. and Meadows, D. H. (2004) Limits to Growth: The 30-Year Update, White River Junction, VT: Chelsea Green Publishing Company.

Ministers of African States (2001) African Ministerial Statement to the World Summit on Sustainable Development, adopted at the African Preparatory Conference for the World Summit on Sustainable Development, 15–18 October 2001, Nairobi.

MMWR (2011) 'Vitalsigns: Current cigarette smoking among adults over 18, United States, 2005–2010', Morbidity and Mortality Weekly Report, 60(35). Available online at http://www.cdc.gov/mmwr/pdf/wk/mm6035.pdf#page=21

Moldan, B. (2011) in correspondence with the authors

Mulenkei, L. (2010) 'If words were action, we would never worry', Outreach, 13 May, London: Stakeholder Forum.

National Academy of Science (1992) Policy Implications of Greenhouse Warming, Washington DC, National Academy Press. Available online at http://www.nap.edu/openbook. php?record_id=1605&page=503

Novo, G. and Murphy, C. (2001) 'Urban agriculture in the city of Havana: A popular response to a crisis', in N. Bakker, M. Dubbeling, S. Guendel, U. Sabel Koschella and H. de Zeeuw (eds.) Growing Cities Growing Food: Urban Agriculture on the Policy Agenda: A Reader on Urban Agriculture, Leusden, Netherlands: Resource Centers on Urban Agriculture and Food Security. Available online at http://www.ruaf.org/node/82

NRG4SD (2002) 'Gauteng Declaration', Brussels: Network for Regional Government for Sustainable Development.

OECD (2011) All ODA figures are from the OECD statistics website - stats.oecd.org/qwids/

OECD (no date) Official development assistance data from www.oecd.org/dac/stats/data

Osborn, D. (2000) 'Introduction', in F. Dodds (ed.) Earth Summit 2002: A New Deal, London: Earthscan.

Osborn, D. and Bigg, T. (1998) Earth Summit II: Outcomes and Analysis, London: Earthscan.

Outreach (2005) 'Stakeholder views on participation at the CSD', Outreach, 15 April, Stakeholder Forum, London.

Pernick, R., Wilder, C. and Winnie, T. (2010) Clean Tech Job Trends 2010, San Francisco, CA: Clean Tech. Available online at http://www.cleanedge.com/sites/default/files/reports/JobTrends2010.pdf

Prahalad, C. K. (2005) The Fortune at the Bottom of the Pyramid, UpperSaddle River, NJ : Prentice Hall.

Prinet, E. (2011) Sustainable Consumption and Production, Vancouver: One Earth Initiative Society.

Puddington, A. (2011) Freedom in the World 2011, Washington, DC: Freedom House. Available online at http://www.freedomhouse.org/report/freedom-world/freedom-world-2011

Raworth, K. (2011) 'From planetary ceilings to social floors: can we live inside the doughnut?', Oxfam. Available online at http://www.oxfamblogs.org/fp2p/?p=7237

Reilly, W. (2011) 'In the GOP's crackpot agenda, *Rolling Stone*, 7 December.

Representatives of Local Governments (2002) Local Government Declaration to the World Summit on Sustainable Development, World Summit on Sustainable Development, 26 August–4 September 2002, Johannesburg, South Africa.

Roan, L. S. (1989) *Ozone Crisis*, New York: Wiley.

Robbins, T. (2003) 'A chill wind is blowing in this nation', speech given to the National Press Club, Washington, DC, April 15. Available online at http://www.commondreams.org/views03/0416-01.htm

Rockström, R. et al. (2009) 'Planetary boundaries: exploring the safe operating space for humanity', *Ecology and Society* 14(2): 32.

Roczniak, D. (1998) Speech by representative of the US chemical industry, the Chemical Manufacturers Association at UNCSD Multi-Stakeholder Dialogue, 12 March 1998.

Royer, L. (1996) 'Speech to the UN Commission on Sustainable Development', in United Nations Commission on Sustainable Development, *Three Years since the Rio Summit*, London: Stakeholder Forum.

Sachs, J. (2005) *The End of Poverty*, London: Penguin. Available online at http://www.earth.columbia.edu/pages/endofpoverty/oda

Sands, P. (1997) 'The Rio Declaration', in F. Dodds (ed.) *The Way Forward Beyond Agenda 21*, London: Earthscan.

Sarkozy, N. (2009) 'France's statement supporting Rio+20'. Available online at www.stakeholderforum.org/fileadmin/files/France%20Statement%20Supporting%20Rio+20.pdf

Schumacher, E. F. (1973) *Small Is Beautiful: A Study of Economics As If People Mattered*, New York: Harper & Row.

Seaford, C., Mohony, S., Wakernagel, M., Larson, J. and Gallegos, R. (2011) *Beyond GDP: Measuring Our Progress*, Global Transition 2012, paper 2, London: Stakeholder Forum.

Sherman, R. and Dodds, F. (2010) *International Governance for Sustainable Development and Rio+20*, London: Stakeholder Forum.

Sherman, R., Peer, J., Dodds, F. and Küpcü Figuera, M. (2005) *Strengthening the Johannesburg Implementation Track*, London: Stakeholder Forum.

Simmons, M. (2009) 'Dennis Meadows wins Japan Prize'. Available online at http://peakenergy.blogspot.com/2009/01/dennis-meadows-wins-japan-prize.html

Simms, A. (2009) 'A green new deal', in F. Dodds (ed.) *Climate and Energy Insecurity*, London: Earthscan.

SIWI (2005) *Making Water a Part of Economic Development: The Economic Benefits of Improved Water Management and Services*, Stockholm: Stockholm International Water Institute.

Stakeholder Forum (2000) 'Non paper on WSSD', London: Stakeholder Forum. Available online at http://www.stakeholderforum.org/fileadmin/files/Earth%20Summit%202012new/WSSD/Non%20Paper%202%20February%202000.pdf

Stakeholder Forum (2009) *Toward a World Summit on Sustainable Development in 2012*, London: Stakeholder Forum. Available online at http://www.stakeholderforum.org/fileadmin/files/IGSD%20Discussion%20Paper%201.pdf

Stakeholder Forum (2011a) *Principles for a Green Economy*, London: Stakeholder Forum, Earth Charter Institute and Bio-regional, London. Available online at www.stakeholderforum.org/fileadmin/files/Principles%20FINAL%20LAYOUT.pdf

Stakeholder Forum (2011b) *Review of Agenda 21 and the Rio Principles*, London: Stakeholder Forum.

Stakeholder Forum, ANPED, Brazilian Forum NGOs and Social Movements for Environment and Development and NGLS (2007a) *Options for Strengthening the Environmental Pillar of Sustainable Development: Compilation of Civil Society Proposals on the Institutional Framework for the United Nations*, London: Stakeholder Forum.

Stakeholder Forum, ANPED, Brazilian Forum NGOs and Social Movements for Environment and Development and NGLS (2007b) *The UN System and Sustainable Development: Proposals for a Sustainable Development Institutional Initiative*, London: Stakeholder Forum.

Steiner, A. (2009) 'Japan and the Republic of Korea launch green new deals'. Nairobi: UNEP. Available online at http://www.unep.org/GreenEconomy/InformationMaterials/News/PressRelease/tabid/4612/language/n-US/Default.aspx?DocumentId=556&ArticleId=6035

Stevenson, A. (1965) Speech to the United Nations, Geneva, 9 July. Available online at http://www.brainyquote.com/quotes/keywords/spaceship.html#ixzz1iRykECKa

Stiglitz, J. (2009) 'Ties to Wall Street doom bank rescue', *Bloomberg News*, 17 April. Available online at http://www.bloomberg.com/apps/news?pid=20601087&sid=afYsmJyngAXQ&refer=home

Stigson, B. (1998) 'How much can be left to the Private Sector and the Market?', speech by WBCSD President, 11 March 1998.

Stoddart, H. (1971) *The Founex Report on Development and Environment*. Available online at http://www.stakeholderforum.org/fileadmin/files/Earth%20Summit%202012new/Publications%20and%20Reports/founex%20report%201972.pdf

Strandenaes, J.-G. (2011) *A Council for Sustainable Development: A Possible Outcome of the Rio+20 Process*, London: Stakeholder Forum.

Strauss, M. (ed.) (1988) *The Dialogue Records. Government-Business-Labor-NGO Interaction at the UN CSD, Year One*, Northern Clearinghouse.

Strong, M. (1972) Opening Speech to the Stockholm Conference on Human Environment, Stockholm, 16 June. Available online at http://www.mauricestrong.net/speeches2/speeches2/Page-10.html

Strong, M. (1992) Available online at http://www.mauricestrong.net/20081004165/rio/rio/rio2.html

Strong, M. (2001) *Where on Earth are We Going?*, Toronto: Vintage Canada.

Suzuki, S. (1992) Information needed. Available online at http://www.4tgc.org/content/page/severin-suzuki-1992

Taibbi, M. (2011) 'Why isn't Wall Street in jail?' *Rolling Stone*, 3 March. Available online at http://www.rollingstone.com/politics/news/why-isnt-wall-street-in-jail-20110216

Tanka, N. (2009) 'Energy efficiency and renewable energy – A key to a better tomorrow', International Energy Agency, ISO Open Session, Cape Town, 17 September. Available online at at www.iea.org/speech/2009/Tanaka/ISO_speech.pdf

Tea Party (2011) 'Tea Party Report on Agenda 21'. Available online at http://www.teapartytribune.com/2011/07/02/agenda-21-conspiracy-theory-or-real-threat/

Thant, U. (1970) Speech on Earth Day. Available online at http://en.wikipedia.org/wiki/Earth_Day

Töpfer, K. (2001) 'Opening statement to European Regional Preparatory Meeting', 24 September. Available online at http://www.unece.org/press/pr2001/01env12e.htm

Toulmin, C. (1997) 'The Desertification Convention', in F. Dodds (ed.) *The Way Forward Beyond Agenda 21*, London: Earthscan.

Trends in Sustainable Development (2008) Available online at http://www.un.org/esa/sustdev/publications/trends2008/fullreport.pdf

Trends in Sustainable Development – Africa (2008) Available online at http://www.un.org/esa/sustdev/publications/trends_africa2008/fullreport.pdf

Turner, G. (2008) *A Comparison of 'The Limits to Growth' with Thirty Years of Reality*, Canberra: CSIRO.

Tutu, D. (2008) Speech at Ubuntu Women Institute USA (UWIU). Available online at http://en.wikipedia.org/wiki/Ubuntu_(philosophy)

UNCSD (2003) Available online at http://daccess-dds-ny.un.org/doc/UNDOC/GEN/N03/384/79/PDF/N0338479.pdf?OpenElement

UNCSD (United Nations Commission on Sustainable Development) (1995) *Three Years Since the Rio Summit*, London: Stakeholder Forum.

UNCSD (1999) Decision 7/3: Tourism and sustainable development, Commission on Sustainable Development, 7th session, New York: United Nations. Available online at http://www.un.org/esa/sustdev/sdissues/tourism/tourism_decisions.htm

UNCSD (2011) Submission of Mexico to the Rio+20 Conference. Available online at http://www.uncsd2012.org/rio20/index.php?page=view&type=510&nr=226&menu=115

UNCSD Secretariat (2011) IEG Consultation on IFSD (3 June 2011) Brief 9, UNCSD, New York. Available online at http://www.uncsd2012.org/rio20/content/documents/SOLO%20DISCUSSION%20PAPER_TEXT.pdf

UNCTC (1991) *Transnational Corporations and Sustainable Development*, Geneva: United Nations. Available online at http://unctc.unctad.org/data/ec1019922a.pdf

UN DESA (2002) 'Key outcomes of the summit'. Available online at http://www.johannesburgsummit.org/html/documents/summit_docs/2009_keyoutcomes_commitments.pdf

UN DESA (2011) Discussion Paper, New York: UN DESA. Available online at http://www.uncsd2012.org/rio20/content/documents/SOLO%20DISCUSSION%20PAPER_TEXT.pdf

UN DPI (2010) 'UN Summit concludes with adoption of global action plan to achieve development goals by 2015', New York: United Nations. Available online at http://www.un.org/en/mdg/summit2010/pdf/Closing%20press%20release%20FINAL-FINAL%20Rev3.pdf

UNECE (1998) Convention on Access to Information, Public Participation in Decision-making and Access to Justice in Environmental Matters (The Aarhus Convention). Available online at http://www.unece.org/env/pp/introduction.html

UNEP (1982a) Report of the Governing Council (Session of Special Character and Tenth Session), New York: United Nations. Available online at http://www.unep.org/resources/gov/prev_docs/82_05_GC10_%20special_character_report_of_the_GC_10_1982.pdf

UNEP (1982b) Nairobi Declaration, Nairobi: UNEP. Available online at http://www.unep.org/Law/PDF/NairobiDeclaration1982.pdf

UNEP (1999) *Global Environment Outlook 2000*, Nairobi: UNEP. Available online at http://www.unep.org/Geo2000/

UNEP (2008) *Green Jobs: Towards Decent Work in a Sustainable, Low-Carbon World*, Nairobi: UNEP. Available online at http://www.unep.org/labour_environment/PDFs/Greenjobs/UNEP-Green-Jobs-Report.pdf

UNEP (2009) International Environmental Governance Reform Process Consultative Group on IEG – 2009, Implementation of Governing Council Decision 25/4, Nairobi: UNEP. Available online at http://www.unep.org/environmentalgovernance/IEGReform/tabid/2227/language/en-US/Default.aspx

UNEP (2010) Co-Chairs of the Consultative Group Summary, Building on the Set of Options for Improving International Environmental Governance of the Belgrade Process, Nairobi: UNEP.

UNEP (2011a) *Towards a Green Economy: Pathways to Sustainable Development and Poverty Eradication*, Nairobi: UNEP. Available online at http://www.unep.org/greeneconomy/ GreenEconomyReport/tabid/1375/Default.aspx

UNEP (2011b) 'Input to the compilation document for UNCSD', Nairobi: UNEP. Available online at http://www.uncsd2012.org/rio20/index.php?page=view&type=510&nr=217 &menu=20

UNEP FI Asset Management Working Group (2007) *Unlocking Value: The Scope for Environmental, Social and Governance Issues in Private Banking*, Geneva: UNEP FI Asset Management Working Group.

UNFCCC (2007) *Investment and Financial Flows to Address Climate Change*, Bonn: United Nation Framework Convention on Climate Change,

UN-Habitat (2011) *Cities and Climate Change: Global Report on Human Settlements 2011*, London: Earthscan.

United Nations (1972) Stockholm Declaration. A/Conf.48/14/Rev. 1 (1973); 11 I.L.M. 1416 (1972) New York: United Nations. Available online at http://www.unep.org/Documents. Multilingual/Default.Print.asp?DocumentID=97&ArticleID=1503&l=en

United Nations (1974) Declaration for the Establishment of a New International Economic Order, UN General Assembly Resolution A/RES/S-6/3201. New York: United Nations. Available online at htpp:// http://www.un-documents.net/s6r3201.htm

United Nations (1983) UN General Assembly Resolution 38/161, New York: United Nations.

United Nations (1992a) Agenda 21: *The United Nations Programme of Action from Rio*, New York: United Nations. Available online at http://www.un.org/esa/dsd/agenda21/res_ agenda21_00.shtm l

United Nations (1992b) Institutional Arrangements to Follow Up the United Nations Conference on Environment and Development, UN General Assembly Resolution A/47/719, New York: United Nations. Available online at http://www.un.org/ documents/ga/res/47/ares47-191.htm

United Nations (1992c) Rio Declaration, New York: United Nations.

United Nations (1993) Institutional Arrangements for the Follow Up to United Nations Conference on Environment and Development, New York: United Nations.

United Nations (1995) *Our Global Neighbourhood: The Report of the Commission on Global Governance*, New York: United Nations.

United Nations (1998) 'Environment and human settlements, Report of the Secretary-General', A/53/463, General Assembly Fifty-third session, New York: United Nations. Available online at http://www.un.org/documents/ga/docs/53/plenary/a53-463.htm

United Nations (2000) Ten-year review of progress achieved in the implementation of the outcome of the United Nations Conference on Environment and Development. General Assembly Resolution A/55/582/Add.1, New York: United Nations.

United Nations (2001) 'Phnom Penh Regional Platform on Sustainable Development for Asia and The Pacific', New York: United Nations. Available online at http://www. johannesburgsummit.org/html/prep_process/asiapacific.html

United Nations (2002a) 'Rio de Janeiro Platform for Action on the Road to Johannesburg, 2002', UN Economic Commission for Latin America and the Caribbean, New York:

United Nations. Available online at http://www.johannesburgsummit.org/html/documents/un_docs/pc2doc5add2e_rio_dejaneiro.pdf

United Nations (2002b) Johannesburg Plan of Implementation, New York: United Nations. Available online at http://www.un.org/esa/sustdev/documents/WSSD_POI_PD/English/POIToc.htm

United Nations (2005) World Summit Outcome. General Assembly Resolution 60/1. 2005, New York: United Nations.

United Nations (2006) *High Level Panel on System Wide Coherence Consultation with Civil Society*, New York: United Nations.

United Nations (2009) 'Implementation of Agenda 21, the Programme for Further Implementation of Agenda 21 and the outcomes of the World Summit on Sustainable Development', A/64/420/Add.1, New York: United Nations. Available online at http://www.un.org/en/ga/64/resolutions.shtml

United Nations (2010) 'Foreword by Ban Ki-moon', in *Millennium Development Goals Report 2010*, New York: United Nations.

United Nations (2011) 'Adopting resolution on multilingualism, General Assembly emphasizes importance of equality among six official United Nations languages', Sixty-fifth General Assembly Plenary, 109th Meeting, New York: United Nations. Available online at http://www.un.org/News/Press/docs/2011/ga11116.doc.htm

United Nations General Assembly (1992d) UN General Assembly resolution A/RES/47/191, New York: United Nations

UN NGLS (1997) *Implementing Agenda 21*, New York: United Nations.

United Nations Secretary-General (1997) *Renewing the United Nations: A Programme for Reform*. Available online at http://www.undp.se/assets/Ovriga-publikationer/Renewing-the-United-Nations.pdf

University of Cambridge Programme for Sustainability Leadership (2012) *The Leadership Compact: Committing to Natural Capital*, Cambridge: University of Cambridge.

Upton, S. (2001) *Sustainable Development Nine Years after Rio*, Paris: OECD.

US Congress (1997) 'Byrd–Hagel Resolution', 105th Congress, 1st Session, Special Resolution 98 sponsored by Senator Robert Byrd (D-WV) and Senator Chuck Hagel (R-NE), 25 July. Available online at http://www.nationalcenter.org/KyotoSenate.html

van Schalkwyk, M. (2009), 'Speech to UNEP Governing Council', UNEP, Nairobi, 19 February.

Ward, B. and Dubos, R. (1972) *Only One Earth: The Care and Maintenance of a Small Planet*, New York and London: W.W. Norton & Company.

WCED (1987) *Our Common Future*, New York: United Nations.

Weissman, R. (1992) 'Summit games: Bush busts UNCED', *Multinational Monitor*, July/August. Available online at http://multinationalmonitor.org/hyper/issues/1992/07/mm0792_06.html

Wikipedia (n.d.) 'Montreal Protocol'. Available online at http://en.wikipedia.org/wiki/Montreal_Protocol

Woodward, D. and Simms, A. (2006) *Growth is Failing the Poor: The Unbalanced Distribution of the Benefits and Costs of Global Economic Growth*, Working Paper 20, New York: UN DESA.

World Economic Forum (2010) *Global Risks 2010: A Global Risk Network Report*, Geneva: World Economic Forum.

World Resources Institute (2011) *Access Initiative*, Washington DC: WRI.

WWF (n.d.) *One Planet Living*. Available online at http://wwf.panda.org/what_we_do/how_we_work/conservation/one planet_living/

Zadek, S. (2011) 'Global finance and banking', Global Transition 2012, Challenge Papers. Available online at http://globaltransition2012.org/challenge-papers/global-financing-and-banking/

Zukang, S. (2008) 'New sustainable development trends reports show mix of promise and danger for agriculture and food security', New York: United Nations. Available online at http://www.un.org/esa/sustdev/csd/csd16/documents/pressrelease_trends.pdf

Index